Energy in World History

ESSAYS IN WORLD HISTORY

William H. McNeill and Ross E. Dunn, *Series Editors*

Energy in World History

Vaclav Smil

University of Manitoba

Westview Press
Boulder • San Francisco • Oxford

Essays in World History

Published in 1994 in the United States of America by Westview Press, Inc., 5500 Central Avenue, Boulder, Colorado, 80301-2877, and in the United Kingdom by Westview Press, 36 Lonsdale Road, Summertown, Oxford OX2 7EW

Library of Congress Cataloging-in-Publication Data
Smil, Vaclav.
    Energy in world history / Vaclav Smil.
      p.   cm.
    Includes bibliographical references and index.
    ISBN 0-8133-1901-3. — ISBN 0-8133-1902-1 (pbk.)
    1. Power resources—History.   I. Title.
TJ163.5.S623   1994
333.79'09—dc20   94-16485
                                                                                                          CIP

Printed and bound in the United States of America

    The paper used in this publication meets the requirements
              of the American National Standard for Permanence of Paper
              for Printed Library Materials Z39.48-1984.

10   9   8   7   6   5   4   3   2   1

Every event in history can occur only insofar as there is available whatever amount of energy (i.e., work) is necessary to carry it out. We can think thoughts wildly, but if we do not have the wherewithal to convert them into action, they will remain thoughts. … History acts in unpredictable ways. Events in history, however, necessarily take on a structure or organization that must accord with their energetic components.

—Richard N. Adams,
Paradoxical Harvest (1982)

Contents

4   Preindustrial Prime Movers and Fuels   92

5   Fossil-Fueled Civilization   157

Illustrations

Figures

Tabular Appendixes

Acknowledgments

This book came about through a combination of old interests and new, chance encounters. I have been fascinated by the nature and development of energy systems ever since my undergraduate years. Although most of my research since then has been devoted to contemporary energy matters (and during the 1970s I also did a good deal of modeling and technical forecasting), my preference for interdisciplinary studies has led me to look repeatedly at the historical developments and implications of energy use and environmental and economic change. I eventually systematized much of this historical work in *General Energetics,* a broad review and analysis of all important aspects of energy flows in the biosphere and their uses by mankind.

In April 1991, just as *General Energetics* was being published, I came across an announcement of a new book series on world history edited for Westview Press by William H. McNeill and Ross E. Dunn. Before I finished reading the paragraph I wanted to do a book on energy in world history. This encounter must be ascribed to Westview's persistence in sending out annual catalogs. In 1982 I had written a book for the house, and they had kept me on their mailing list. Even more fortuitously, I had met Professor McNeill in October 1987 at the symposium on The Earth as Transformed by Human Action, and I had never forgotten the meandering conversation we had while walking through the Harvard Forest at Petersham. I wrote to him immediately after seeing the announcement for the new series, hoping he would like my proposal.

In two weeks I had his encouraging letter, and setting aside all other work, I started to write a sample chapter to test the pitch and style of the book. In summer 1991 I had to return to my abandoned obligations: *Global Ecology* and several smaller environmental and energy projects kept me busy for a year. In fall 1992 I returned to the book, and by summer 1993 the writing was completed. Professor McNeill's criticism and suggestions at every stage of the process helped to make the book not only clearer and more accurate but also richer.

Drafts were also read by David Keith, an MIT-trained physicist with an interest in interdisciplinary energy matters, and David Smil, my son. They watched for scientific accuracy and for the clarity of technical terms and explanations. Two graphic artists who have worked on most of my previous projects, Marjorie Halmarson and Ed Pachanuk, prepared the numerous reproductions of historical illustrations and many original images and graphs. Once again, their work has helped to enrich my own.

Acknowledgments

Obviously, my greatest debt is to hundreds of historians, scientists, engineers, and economists. Without their research this book would not have been possible. I owe them also many thanks for a great learning experience.

Vaclav Smil

1

Energy and Society

Obviously, if we can find a single word to represent an idea which applies to every element in our existence in a way that makes us feel we have a genuine grasp of it, we have achieved something economical and powerful. This is what has happened with the idea expressed by the word energy. No other concept has so unified our understanding of experience.

—R. Bruce Lindsay, *Energy* (1975)

ENERGY IS THE ONLY universal currency: It must be transformed to get anything done. Manifestations of these transformations in the physical universe range from rotating galaxies to the erosive forces of tiny raindrops. Life on Earth, the only known life in the universe, would be impossible without the photosynthetic conversion of solar energy into plant biomass. Humans depend on this transformation for their survival and on many more energy flows for their civilized existence.

The evolution of human societies has been dependent upon the conversion of ever larger amounts of ever more concentrated and more versatile forms of energy. From the perspective of natural science, both prehistoric human evolution and the course of history may be seen fundamentally as the quest for controlling greater energy stores and flows. This endeavor has brought about the expansion of human populations and allowed for increasingly complex social and productive arrangements. Neither the growth of technical capabilities and a deeper understanding of the surrounding world nor the effort to secure a better quality of life would have been successful without innovations in energy use.

As formulated by Alfred Lotka (1925) in his law of maximum energy, natural selection will tend to increase the total mass of an organic system, and this will increase the rate of circulation of matter as well as the total energy flux through the system—as long as there is a surplus of available energy. The history of successive civilizations, the largest and most complex organisms in the biosphere, has followed this course. Human dependence on ever higher energy flows can be seen as an inevitable continuation of organismic evolution.

Starting with Wilhelm Ostwald (1909), a Nobel prize–winning chemist, twentieth-century scholars have repeatedly made the link between energy and civilization. Two quotations will suffice to illustrate this relationship. In a pioneering paper, anthropologist Leslie White (1943) called the link the first important law of cultural development: "Other things being equal, the degree of cultural development varies directly as the amount of energy per capita per year harnessed and put to work" (p. 338). Two generations later, a physicist, Ronald E. Fox (1988), concluded a book on energy in evolution by writing, "A refinement in cultural mechanisms has occurred with every refinement of energy flux coupling" (p. 166). Assessing the validity of such conclusions is a major goal of this book.

But even Nobel prize winners encounter great difficulties when they try to give a satisfactory answer to a seemingly simple question: What is energy? Richard Feynman (1988), one of the greatest physicists of the twentieth century, stressed, "It is important to realize that in physics today, we have no knowledge of what energy *is*. We do not have a picture that energy comes in little blobs of a definite amount" (p. 4–2). What we do know is that all matter is energy at rest, that energy manifests itself in a multitude of ways, and that these distinct energy forms are linked by numerous conversions (see Figure 1.1). Scientists rapidly expanded and systematized their understanding of these stores, potentials, and transformations during the nineteenth century and perfected it during the twentieth; surprisingly, they discovered how to release nuclear energy in the late 1930s, two decades before they understood how photosynthesis works.

Flows, Stores, and Controls

The availability of power sources determines the amount of work activity that can exist, and control of these power flows determines the power in man's affairs and in his relative influence on nature.

—Howard T. Odum, *Environment, Power, and Society* (1971)

All known forms of energy are critical for human existence. This reality precludes any rank-ordering of their importance. Much in the course of history has been determined and circumscribed by universal and planetary flows of energy and their regional or local manifestations. Fundamental features of the universe are governed by **gravitational energy**, which orders countless galaxies and star sets akin to our solar system. Gravity also keeps our planet orbiting at just the right distance from the sun and holds the atmosphere, which makes the Earth habitable (see A1.1).

Like all active stars, the sun is powered by **nuclear energy.** The product of those thermonuclear reactions reaches the Earth in the form of **electromagnetic (solar, or radiant) energy.** The flux of this energy ranges over a broad spectrum of wavelengths, including visible light, and about a third of this huge flow is reflected by clouds and surfaces. Nearly all of the remainder is absorbed by oceans, land, and

TO \ FROM	ELECTRO-MAGNETIC	CHEMICAL	NUCLEAR	THERMAL	KINETIC	ELECTRICAL
ELECTRO-MAGNETIC		CHEMILUMINES-CENCE	NUCLEAR BOMBS	THERMAL RADIATION	ACCELERATING CHARGES	ELECTRO-MAGNETIC RADIATION
CHEMICAL	PHOTO-SYNTHESIS	CHEMICAL PROCESSING		BOILING	DISSOCIATION BY RADIOLYSIS	ELECTROLYSIS
NUCLEAR	GAMMA-NEUTRON REACTIONS					
THERMAL	SOLAR ABSORPTION	COMBUSTION	FISSION / FUSION	HEAT EXCHANGE	FRICTION	RESISTANCE HEATING
KINETIC	RADIOMETERS	METABOLISM	RADIOACTIVITY / NUCLEAR BOMBS	THERMAL EXPANSION / INTERNAL COMBUSTION	GEARS	ELECTRIC MOTORS
ELECTRICAL	SOLAR CELLS	FUEL CELLS / BATTERIES	NUCLEAR BATTERIES	THERMO-ELECTRICITY	ELECTRICITY GENERATORS	

FIGURE 1.1 Matrix of energy conversions. Where more possibilities exist, only one or two leading transformations are identified.

the atmosphere, converted to **thermal energy,** and reradiated by the planet. **Geo-thermal energy** results from the original gravitational accretion of the Earth's planetary mass and from the decay of radioactive matter. These flows drive the grand tectonic processes that continually reorder the oceans and continents and cause volcanic eruptions and earthquakes.

Only a tiny part of radiant energy is transformed by photosynthesis into new stores of **chemical energy** in plants. These stores provide the irreplaceable foundation for all higher life. Animate metabolism reorganizes nutrients into growing tissues and maintains body functions and, in all mammals, also constant body temperature. Digestion also generates **mechanical (kinetic) energy** of working muscles. In their energy conversions animals deploy their muscles in search of food, reproduction, escape, and defense, but these functions are limited by the size of their bodies and by the availability of accessible nutrition.

Humans can extend these physical limits by using tools and harnessing the energies outside their own bodies. Unlocked by the human intellect, these extra-somatic energies have been used for a growing variety of tasks; they serve both as powerful **prime movers** and as **fuels** that release heat during combustion. Two conditions must be satisfied in order for humans to convert these energies to their own uses. First, the requisite energy flows (water, wind) or potentials (animals, biomass, fossil or nuclear fuels) must be present in exploitable quantities. Second, humans must take the actions or deploy the controls necessary to capture or release these flows and potentials in useful forms. The triggers of energy supplies depend on the flow of information and on an enormous variety of artifacts.

These devices have ranged from such simple tools as hammerstones and levers to complex fuel-burning engines and reactors that can release the energy of nuclear fission. The basic evolutionary and historical sequence of these advances is easy to outline in broad qualitative terms. Like any nonphotosynthesizing organism, humans require food. This is their most fundamental energy need. Foraging and scavenging by early hominids was very similar to the food acquisition strategies of their primate ancestors (Whiten and Widdowson 1992). Although some primates have a rudimentary tool-making capability, only hominids have explored this potential in a sustained manner.

Tools have given people many mechanical advantages in the provision of food, shelter, and clothing. The mastery of fire greatly extended humanity's range of habitation and set humans further apart from animals (Goudsblom 1992). Later, the invention of better tools allowed people to harness domesticated animals, build complex muscle-powered machines, and convert a tiny fraction of the huge kinetic energies of wind and water into useful mechanical power.

These new prime movers greatly enlarged useful power under human command, but for a very long time their effective use was circumscribed by the nature and magnitude of the captured flows. Most obviously, this was the case with sailing. Solar energy inputs govern the basic patterns of atmospheric and oceanic circulation. Prevailing wind flows and persistent ocean currents are fashioned by location and the interaction between land and water masses. These grand flows steered the late fifteenth-century European transatlantic voyages to the Caribbean and prevented the Spaniards from discovering Hawaii even after nearly three centuries of sailing through the Pacific.

The development of controlled combustion in fireplaces, stoves, and furnaces enabled people to turn the chemical energy of plants into thermal energy. Society began to use this heat not only directly in households but also for commercial purposes. It enabled them to smelt metals, to fire bricks, and to process and finish countless products. Combustion of fossil fuels made all of these traditional uses of heat more widespread and more efficient. A number of fundamental inventions made it possible to convert thermal energy produced from the burning of fossil fuels to mechanical energy. Steam and internal combustion engines were the earliest of these inventions; gas turbines and rockets followed. Since the end of the nineteenth century all modernizing societies have been using fossil fuels as well as the kinetic energies of water and wind to generate **electrical energy**. Since the 1950s an increasing number of nations has also been generating electricity by using nuclear energy released by the fissioning of heavy atoms. Fossil fuels and electricity have created a new form of high-energy civilization that has quickly expanded to encompass the whole planet.

What is much more difficult than outlining this grand sequence is to set these developments in a broader perspective. This attempt requires a number of

approaches—some straightforward and others more complex. It is relatively easy
to evaluate the advantages and drawbacks of the more potent prime movers and
the better fuels. It is more difficult to identify the factors that promote or impede
innovation—that is, the processes that enable a society to take the intellectual and
technical steps needed to unlock great energy potentials. As these changes take
place, they have important consequences for farming, industry, transport, settlement patterns, warfare, and the Earth's environment. An appraisal of these impacts is equally complex. But no serious attempt to address these matters could be
solely qualitative. Quantitative accounts are essential in order to appreciate not
only the magnitude and the limits of human achievements but also their consequences. Naturally, their comprehension requires knowledge of basic concepts
and measures.

Concepts and Measures

The key idea is simple: constancy in the midst of change.
—R. Bruce Lindsay, *Energy* (1975)

A number of first principles underlies all energy conversions. Every form of energy can be turned into heat, or thermal energy. No energy is ever lost in any of
these conversions. The **conservation of energy** is one of the most fundamental
universal realities. But as the conversion chain progresses, the potential for useful
work steadily diminishes (see A1.2). Physicists call the measure associated with
this loss of useful energy **entropy.** Although the energy content of the universe is
constant, conversions can only increase the entropy of a closed system, that is, decrease its utility. There is nothing we can do about this inexorable reality. A bas-
ketful of grain or a barrelful of crude oil are low-entropy stores of energy; they are
capable of much useful work once metabolized or burned. The random motion of
slightly heated air molecules is an irreversible high-entropy state representing an
irretrievable loss of utility.

This unidirectional entropic dissipation leads to a loss of complexity and to
greater disorder and homogeneity in any closed system. But living organisms—
from the smallest bacteria to the largest civilizations—temporarily defy this trend
by importing and metabolizing energy. This means that every living organism is
an open system maintaining continuous inflow and outflow of energy and matter.
As long as organisms are alive, they can never be in a state of chemical and thermodynamic equilibrium (von Bertalanffy 1968). Their growth and evolution result in greater heterogeneity and in increasing structural and systemic complexity.

Scientists did not fully understand these realities until the nineteenth century.
Before 1850, those immersed in the rapidly evolving disciplines of physics, chemistry, and biology found a common concern in studying transformations of en-

FIGURE 1.2 Two eighteenth-century horses turning a capstan geared to pumping well water used for dying carpets in a French manufacture. Typical horses of this period could not sustain a steady work rate of 1 horsepower. James Watt used an exaggerated rating in order to assure customers' satisfaction with his horsepower-denominated steam engines. *Source:* Reproduced from Diderot and D'Alembert (1769–1772).

ergy (Cardwell 1971; Lindsay 1975). These fundamental interests required a codification of standard measurements. The two options that became common for measuring **energy** were the **calorie,** a metric unit, and the **British thermal unit** (see A1.3). Today's basic international unit of energy is the **joule,** named after an English physicist, James Prescott Joule (1818–1889), who published the first accurate calculation of the equivalence of work and heat.

Energy units

Power simply denotes the rate of energy flow. Its first standard unit, the **horsepower,** was set by James Watt. He wanted to charge for his steam engines on a readily understandable basis, and so he chose to compare them with the prime mover they were designed to replace—a harnessed workhorse powering a mill or a pump. But it appears that he did so in a somewhat devious manner (see A1.3 and Figure 1.2). Today's basic scientific unit of power is, appropriately, 1 **watt** (1 joule/second).

Power units

Another important rate is **energy density,** the amount of energy per unit mass of a resource. This value is of critical importance for foodstuffs as well as for fuels. Even where abundant, foods that have a low energy density could never become

Energy density

staples. Pre-Hispanic inhabitants of the Basin of Mexico, for example, always ate plenty of prickly pears, the fruit of the *Opuntia* cactus, which were easy to gather from wild plants (Sanders et al. 1979). But in order for even a small woman to satisfy most of her food energy needs on such a diet, she would have to eat nearly 5 kilograms of the fruit every day. She could get the same amount of energy from just about half a kilogram of tortillas. Conversely, charcoal's high energy density—about twice that of air-dried wood—made it the best fuel for cooking and metallurgy in traditional societies (see A1.4).

Power density, the rate at which energies are produced or consumed per unit of area, constitutes a critical structural determinant of energy systems. For example, in all traditional societies dependent on fuelwood and charcoal, city size was clearly limited by the inherently low power density of biomass production (see A1.5). The power density of sustainable annual tree growth in temperate climates is at best equal to just 1 or 2 percent of the power density of energy consumption required for traditional urban heating, cooking, and manufactures. Consequently, cities had to draw on large areas of the surrounding land—at least fifty to more than one hundred times their size—in order to obtain an adequate fuel supply. This need for fuel restricted their growth even where other resources, like food and water, were abundant.

Yet another rate, one that has assumed much importance with advancing industrialization, is the **efficiency** of energy conversions. This ratio of output to input is used most commonly when describing the performance of energy converters, be they stoves, engines, or lights. Although people cannot do anything about the entropic dissipation of the energy they use, they can try to improve the efficiency of conversions by lowering the amount of energy required to perform specific tasks (see A1.6). Obviously, there are physical limits to these improvements, but in most instances there is still much room for improvement.

When efficiencies are calculated for production of foodstuffs, fuels, or electricity, they are usually called **energy ratios.** Obviously, energy ratios in every prosperous traditional agricultural system had to be greater than one. Edible harvests had to contain more energy than the amount that humans and animals consumed in producing those crops. However, there was no simple relationship between food energy ratios and the social complexity of old high cultures. In contrast, industrial societies prefer to develop the fossil fuel resources with the highest net energy ratios. This is why they favor crude oil in general, and the rich Middle Eastern fields in particular (see A1.7).

Finally, **energy intensity** measures the cost of products or services in standard energy units. Among the commonly used materials, aluminum and silicon are highly energy-intensive, whereas glass and paper are fairly cheap (see A1.8). The technical advances of the past two centuries have brought many substantial declines in energy intensities. One notable example is the coke-fueled smelting of pig iron in large blast furnaces, which needs less than one-tenth the energy per unit mass of finished product that charcoal-based production requires.

Complexities and Caveats

Discoursive thinking always represents only one aspect of ultimate reality. ... It can never exhaust its infinite manifoldness.

—Ludwig von Bertalanffy, *General System Theory* (1968)

Using standard units to measure energy storages and flows is physically straight-forward and scientifically impeccable—yet these reductions to a common denominator mislead in several important ways. They cannot capture critical qualitative differences among various energy sources. Two kinds of coal may have identical energy densities, but one may burn very cleanly and the other may smoke heavily and emit a great deal of sulfur dioxide. The abundance of high-energy-density smokeless coal, the ideal fuel for steam engines, clearly helps to explain why Britain dominated nineteenth-century maritime transport. Neither France nor Germany had comparably good coal resources.

Abstract units obviously cannot differentiate between edible and inedible biomass. Identical masses of wheat and dry wheat straw contain virtually the same amount of heat energy—but the straw cannot be digested by people, whereas wheat is an excellent source of basic nutrients. Neither can the units of measurement reveal the specific characteristics of the food energy, a matter of great importance for proper nutrition. Many high-energy foods contain little or no protein and fat, two nutrients required for normal body growth and maintenance.

There are other important qualities hidden by abstract measures. Access to energy stores is obviously a critical matter. Stemwood and branchwood have the same energy densities, but without good saws, many preindustrial societies could only gather the latter. Conversion efficiency can be decisive in choosing a fuel. Natural gas and fuel oil have similar energy densities, but the best gas furnaces are much more efficient. Ease of use is no less important. Straw burning requires frequent stoking, whereas large wood pieces can burn unattended for hours. Pollution potential makes an enormous difference for both indoor and outdoor air quality. Unvented indoor cooking with dry dung produces much more smoke than the burning of seasoned wood.

Nor does the common denominator of basic measurement differentiate between renewable and fossil energies—yet this distinction is fundamental to a proper understanding of the nature and durability of a given energy system. A perfect illustration of these problems is an often used but basically misleading comparison between the energy ratios of food production in traditional and modern agricultures (see A1.9).

Similar difficulties complicate the use of various rate measures. How powerful and efficient are people as prime movers? The first part of this question was answered quite accurately long before systematic energy studies began in the nineteenth century. The early estimates equated the labor of one horse with the exer-

FIGURE 1.3 Glass polishers working in a French factory. Guillaume Amontons's accurate estimate of sustainable labor was based on the work of glass polishers. They moved, back-and-forth, a pad pressed against the glass by a wooden bow, exerting a horizontal force of just over 10 kilograms at the speed of just under 1 meter per second. In modern units this work rate would be equal to about 90 watts. The polishers had to sustain this rate for 10 hours a day. *Source:* Reproduced from Diderot and D'Alembert (1769–1772).

tion of anywhere from two to fourteen men. By 1699, Guillaume Amontons had made a very good assessment based on the typical effort of glass polishers (Ferguson 1971; see Figure 1.3). Before 1800 a number of values converged on the correct range of from fewer than 70 to slightly more than 100 watts for most steadily working adults. When working steadily at a rate of 75 watts, ten men would be needed to equal the power of one standard horse.

Advances in biochemistry in the late nineteenth century made it possible to determine the highest efficiency of human muscles. During peak aerobic performance, over 20 percent of ingested food energy is actually converted to kinetic energy. But it is still quite difficult to say just how efficient people are as prime movers. Certainly the total daily food intake should not be counted as an energy input of labor: Basal metabolism operates regardless of whether people are working or at rest.

Perhaps a calculation of the net energy cost of human activity provides the most satisfactory solution to this problem (see A1.10). But even in much simpler societies than ours a great deal of labor was mental rather than physical—and the metabolic cost of thinking (even very hard) is small. And yet the development of mental faculties requires years of language acquisition, schooling, experience, and socialization.

A real understanding of energy's role in history requires both quantitative and qualitative evaluations. I have emphasized the need for quantitative appraisals, but one must not reduce everything to numerical accounts and treat them as all-encompassing explanations. Consequently, I will approach the challenge in both ways. I will note energy and power requirements and densities, calculate energy returns as accurately as data permit, and point out improving efficiencies. But I will not forget nonenergetic necessities, motivations, changes, and advances. Without their presence and effects the history of human energy use would have been profoundly different. Energetic imperatives have left a powerful imprint on history. But many details, sequences, and consequences of these fundamental determinants can be explained only by referring to human motivations, longings, and passions, random events, and the sometimes surprising and often inexplicable events of history.

APPENDIXES

A1.1 Gravitation and the Habitability of the Earth

The extreme tolerances of carbon-based life are determined by the freezing point of the liquids required for the formation and movement of organic molecules, on the one hand, and by the high temperatures and pressures that destabilize amino acids and break down proteins, on the other. The continuously habitable zone—the range of orbital radius assuring optimal conditions for a life-supporting planet—is very narrow (Smoluchowski 1983).

If, when the solar system formed, the Earth had settled into an orbit just 5 percent closer to the sun, it is unlikely that life would have evolved on the planet; a strong greenhouse effect would have caused intolerably high temperatures. In contrast, if the planet had taken up a position a mere 1 percent further from the sun, all of its water would have been locked in glaciers. Without the Earth's gravity the planet's unique atmosphere—dominated by nitrogen, enriched by oxygen from photosynthesis, and containing a number of important trace gases regulating surface temperature—could not have supported highly diversified life.

A1.2 Diminishing Utility of Energy

Any energy conversion illustrates the principle of diminishing utility. If an incandescent bulb is now illuminating this page, for example, the electromagnetic energy of the light is equivalent to only a small part of the chemical energy contained in the lump of coal used to generate that light. The bulk of the coal's energy has escaped as heat through a plant chimney, into the cooling water condensing the hot steam, through the wiring during transformation and transmission of the electric current, or through the bulb's coiled filament. And the light reaching the page is either absorbed by it, reflected and absorbed by its surroundings, or reradiated as heat.

What can we do with this diffused heat that warmed the air above the station, along the wires, around the light bulb, above your page? No energy has been lost, but the initially highly useful form was degraded to the point where it has no practical use.

A1.3 Measuring Energy and Power

A joule measures the work accomplished when a force of 1 newton acts over a distance of 1 meter. A less abstract approach is to define the basic energy unit through heat requirements. One calorie is the amount of heat energy needed to increase the temperature of 1 cubic centimeter of water by 1 degree Celsius. This is a tiny amount of energy: Doing the same for 1 kilogram of water calls, naturally, for one thousand times more energy, or 1 kilocalorie (kcal; for the complete list of multiplier prefixes see Basic Measures).

Given the equivalence of heat and work, converting calories to joules is easy: One calorie equals roughly 4.2 joules. For first approximations, simply multiply by four. The conversion is equally simple for the nonmetric English measure of energy, the British thermal unit. One BTU contains roughly 1000 joules (1055 to be exact). A good anchor for energy comparisons is the average daily food consumption. For most people it falls between 2000 and 2700 kcal (2–2.7 megacalories, or Mcal), or about 8–11 megajoules, or MJ.

The first power unit has an interesting history. In 1782 James Watt calculated in his *Blotting and Calculation Book* that a mill horse works at a rate of 32,400 foot-pounds a minute—and the next year he rounded this up to 33,000 foot-pounds (Dickinson 1967). He assumed an average walking speed of about 3 feet per second, clearly a typical performance. But we do not know where he got his figure for an average pull of about 180 pounds. Many large animals working at that time in England were that powerful—but most horses in eighteenth- and even nineteenth-century Europe could not sustain rates of 1 horsepower. Did Watt deliberately overestimate the average power of horses in order not to disappoint buyers of his steam engines?

Today's standard unit of power, 1 watt (W), is equal to the flow of 1 joule per second. One horsepower is equal to about 750 watts. To return to the example of daily food consumption, 8 MJ of food per day corresponds to a power rate of about 90 W (8 MJ divided by the product of 24 hours times 3600 seconds), less than the rating of a standard light bulb (100 W). A double toaster uses electricity at a rate of 1000 W, or 1 kW; small cars consume gasoline at a rate of around 50 kW; and a large coal-fired power plant produces electricity at a rate of 2 gigawatts (GW).

A1.4   Energy Densities of Foodstuffs and Fuels

Foodstuffs and Fuels		Energy Densities (MJ/kg)
Foodstuffs		
Very low	Vegetables, fruits	0.8–2.5
Low	Tubers, milk	2.5–5.0
Medium	Meats	5.0–12.0
High	Cereal and legume grains	12.0–15.0
Very high	Oils, animal fats	25.0–35.0
Fuels		
Very low	Peats, green wood, grasses	5.0–10.0
Low	Crop residues, air-dried wood	12.0–15.0
Medium	Bituminous coals	18.0–25.0
High	Charcoal, anthracites	28.0–32.0
Very high	Crude oils	40.0–44.0

Source: Derived from Smil (1991).

A1.5 Power Densities of Biomass Fuels

Photosynthesis converts on the average less than 1 percent of incoming solar radiation into new biomass. In warm and rainy climates the best sustainable annual fuelwood productivities for fast-growing species are usually no more than 15 tonnes per hectare (t/ha); in drier regions with slower-growing trees, this figure is between 5 and 10 t/ha. With dry wood averaging 18 MJ per kilogram (kg), these rates translate into power densities of 0.3–0.9 W per square meter (m^2). Converting much of this wood into charcoal would have easily halved these power densities.

In the built-up areas of a large preindustrial city in a cold climate, heating, cooking, and producing various manufactures would have required at least 20–30 W per square meter. Consequently, the city would need a nearby area of up to 100 times its size to secure a sustainable supply of fuel.

A1.6 Efficiency Improvements

The cumulative effect of technical innovation and advances in management have often been translated into impressive long-term gains. During the late 1880s, for example, Thomas Edison's first electricity-generating plants converted less than 10 percent of the coal they used to electricity; his first light bulbs turned less than 1 percent of the electricity they used into light. As a result, less than 0.1 percent of the chemical energy in coal ended up as light on a page. Today's best plants are about 40 percent efficient and the best lights about 20 percent efficient: This means that about 8 percent of coal ends up as light, an eighty-fold efficiency gain in a century! Such gains not only save capital and operation costs. More fundamentally, they lower the inevitable environmental impacts: There is less land destroyed by mining, less warm water dumped into streams, and less air pollution.

A1.7 Energy Cost of Energy

Available calculations show the energy cost of U.S. and British coal to be anywhere from 100 kJ/kg to 4 MJ/kg (Boustead and Hancock 1979). The first value implies an efficiency of

99.75 percent, the second one of just over 80 percent. For the richest Middle Eastern oilfields, the energy invested in exploration (less than 1 kJ/kg) and production (0.5–5 kJ/kg) adds up to as little as 0.005 percent of the energy contained in a kilogram of crude oil! This is a negligible total compared to the amount of energy needed for shipping (1–3.5 MJ/kg, including the construction of tankers and oil terminals) and refining (anywhere from 4 to 10 percent of crude oil input). In general, the energy investments needed to produce the fossil fuels that sustain modern civilization produce very large energy returns.

A1.8   Energy Intensities of Common Materials

Material	Energy Cost (MJ/kg)	Remarks
Aluminum	227–342	Metal from bauxite
Bricks	2–5	Baked from clay
Cement	5–9	From raw materials
Copper	60–125	Metal from ore
Explosives	10–70	From raw materials
Glass	18–35	From sand and other materials
Gravel	0.08–0.1	From quarries, rivers
Iron	20–25	From iron ore
Lead	30–50	From ore
Limestone	0.07–0.1	From rock
Paper	25–50	From standing timber
Plastics	60–120	From crude oil
Sand	0.08–0.1	Excavated
Silicon	230–235	From silica
Steel	20–50	Finished from ore
Water	0.001–0.01	From streams, reservoirs
Wood	3–7	From standing timber

Source: Compiled from Smil (1991).

A1.9 Comparison of Energy Ratios in Food Production

Since the early 1970s energy ratios have often been used to illustrate the superiority of traditional farming and the low energy returns of modern agriculture. Such comparisons are misleading owing to a fundamental difference between the two ratios. The figures cited for traditional farming are internally consistent: They are simply quotients of the food energy in crops and the food energy (labor) needed to produce those harvests. In contrast, in modern farming the denominators are composed overwhelmingly of the nonrenewable fossil fuel inputs needed to power field machinery and make the equipment and farm chemicals—and labor inputs are negligible.

If the ratio was calculated merely as a quotient of edible energy output to labor input, then modern systems, with their minuscule amount of human effort and no draft animals, would look superior to any traditional practice. If the cost of producing a modern crop includes all fossil fuels and electricity converted to a common denominator, then the energy returns in modern agriculture fall substantially below traditional returns. Such a calculation is possible because of the physical equivalence of energies: Both food and fuels can be expressed in identical units. But an obvious "apples and oranges" problem remains.

Clearly, there is no satisfactory way to compare—simply and directly—the energy returns of the two fundamentally different farming systems.

A1.10 Calculations of Net Energy Cost of Human Labor

The energy cost of human labor can be expressed in a number of ways, none of them quite satisfactory (Fluck 1992b). I believe that the net energy cost is the most satisfactory choice: This measure quantifies a person's energy consumption above the basic existential need that would have to be satisfied even if the person did no work. This approach debits human labor, with its actual incremental energy cost, rather than exaggerating its value by charging it with the total daily food consumption.

Most of the total food intake is usually claimed by the basal metabolic rate, which covers the maintenance of vital biochemical functions and a steady body temperature. The metabolic rate varies with sex and age and displays considerable individual departures from large-scale group means. To this must be added mark-ups to allow for the energy costs of sitting, standing, and moving in order to eat and carry out basic personal hygiene. Food consumption in excess of these requirements is a rational measure of the energy cost of both labor and leisure activities.

2

Energy in Prehistory

Their lives dragged on, in a mode of beasts that roam
From place to place. And none was there to guide
With sturdy arm the curving plow, and none
Had learned to work the fields with iron, or plant
The earth with tender shoots. ...

—Lucretius, *De rerum natura* (first century b.c.)

THE ORIGINS OF THE genus *Homo* and the details of its subsequent evolution remain controversial. These uncertainties cannot change the fact that hominids have spent all but a small fraction of their existence as simple foragers. For millions of years their natural diet and their foraging strategies resembled those of their primate ancestors (Whiten and Widdowson 1992). Bipedality, datable to about 7.5 million years ago, appears to have been the single most important evolutionary change for hominid predecessors (Leakey and Lewin 1992). This critical trait could have evolved after the tree-dwellers were pushed out from the trees by the more powerful ancestors of today's great apes (Romer 1959), or it could have come as a result of ecosystemic change from a dense forest to an open woodland.

Bipedality started a cascade of enormous evolutionary adjustments (Hockett and Ascher 1964). Walking upright liberated hominid arms for carrying weapons and for taking food to group sites instead of consuming it on the spot. Bipedality also freed the mouth and teeth, which enabled hominids to develop a more complex call system, the prerequisite of language. These developments required larger brains. The energy cost of the human brain eventually reached three times the level for chimpanzees, accounting for up to one-sixth of the total basal metabolic rate (Foley and Lee 1992).

Higher encephalization was critical for the rise of social complexity, which raised the survival odds and set hominids apart from other mammals. But it could not change the energy basis of existence: To secure food hominids relied only on their muscles and on simple stratagems as gatherers, scavengers, hunters, and fishers. It is impossible to trace the genesis of the first wooden tools (sticks and clubs), but a number of East African sites have shown that hominids were

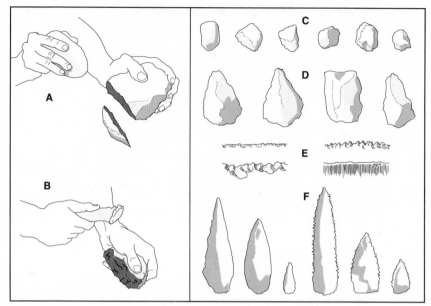

FIGURE 2.1 Stone tools. The first step in making stone tools, the initial reduction of cores, required striking the stone with a suitable hammer stone (*A*). The final reduction was done with softer stones, wood, bone, or antler tools (*B*). The oldest Oldowan tools (*C*) were notably smaller than later Acheulean artifacts (*D*). Great care was taken in preparing special edges (*E*) and fashioning uniface or biface points (*F*). *Sources:* Based on McCarthy (1967), Toth (1987), and Pokotylo (1988).

making stone tools as early as 2.5 million years ago (see Figure 2.1). Cobble-based, these relatively small and simple Oldowan hammerstones, choppers, and flakes made it much easier to butcher animals and break animal bones (Toth 1987).

About 1.5 million years ago hominids started to quarry larger flakes (up to 20 centimeters long) to make hand axes, picks, and cleavers of Acheulean style. These practices were not markedly improved until about 25,000 years ago. The improvements came about with the production of ground and hafted adzes and axes and the more efficient flaking of flint. Using the Magdalenian techniques of 15,000–10,000 B.C., a toolmaker could produce up to 12 meters of microblade edges from a single stone.

Other common artifacts have included spears and bows and arrows (Coles and Higgs 1969). The invention of spear throwers during the late Paleolithic easily doubled the velocity of the weapon and reduced the necessity for a close approach. The stone-tipped arrow carried these advantages further, with the added gain of accuracy.

These societies not only vastly improved their tools and weaponry but also made other important advances. Most notably, they began to control the use of fire for warmth and cooking. Archaeologists disagree about when this advance took place—some postulate that early hominids may have been using fire nearly half a million years ago. More reliable estimates date fire use at about 250,000 years ago (James 1989; Goudsblom 1992).

The great spatial and temporal variability of the preserved archaeological record precludes any simple generalizations concerning the energy balances of prehistoric societies (Freeman 1981). Descriptions of first contacts with surviving foraging societies and modern anthropological studies provide only uncertain analogies. Information on groups that survived in extreme environments long enough to be studied by modern scientific methods offers at best a very limited insight into the lives of prehistoric foragers in more equable climates and more fertile areas. Moreover, the studied societies have often been affected by contacts with pastoralists, farmers, or overseas migrants (Headland and Reid 1989). But the absence of a typical foraging pattern does not preclude the recognition of a number of biophysical imperatives governing energy flows and determining the behavior of gathering and hunting groups.

Foraging Societies

Most of the distinguishing human characteristics, such as bipedality, manual dexterity, and elaborate technology, and marked encephalization can be viewed as having been promoted by the demands of an opportunistic foraging system.
—Erik Trinkaus, in *The Evolution of Human Hunting* (1987)

Most gatherers and hunters did not evolve complex societies. This was largely because they could not attain the population densities necessary for functional specialization and social stratification. Average foraging densities—reflecting a variety of habitats and food acquisition skills and techniques—ranged from just around one person to several hundred people per hundred square kilometers (Murdock 1967; Jochim 1976; Kirk 1981; Kelly 1983). Population densities were commonly about ten times higher in seasonally dry tropical environments than they were in cold climates.

The rates were another order of magnitude higher for groups that had excellent resources for both hunting and gathering (in postglacial Europe and the Basin of Mexico, for example) or that were heavily dependent on aquatic species (in the Baltic region and the Pacific Northwest). Fishing and near-shore hunting of sea mammals sustained the highest foraging densities, led to semi-permanent and even permanent settlements, and sharply reduced a society's food-acquisition mobility on land.

These large density variations were not tied to biospheric energy flows in simple ways. They did not uniformly decrease poleward with lower temperatures.

And they did not increase proportionately to the photosynthetic productivity of ecosystems or to the total mass of animals available for hunting. Any attempt to identify energy imperatives behind the density differences must combine eco-systemic variables with estimates of relative dependence on plant and animal foods and seasonal storage capabilities.

Both gathering and hunting were surprisingly unrewarding in species-rich tropical and temperate forest environments. These ecosystems store most of the planet's vegetation, but they do so mostly in tall tree trunks whose cellulose and lignin humans cannot digest. Energy-rich fruits and seeds are a very small por-tion of total plant mass, and they are mostly inaccessible in high canopies. Seeds are also often protected by hard coats and need a fairly energy-intensive process-ing (shelling, pounding) before consumption. Food gathering in tropical forests may also require extensive roaming: The great variety of species means that there are often considerable distances between the individual trees or vines whose parts may be ready for collection at a given time.

Tropical forests were also the least advantageous ecosystems to exploit by hunt-ing. Most of their animals are arboreal (many also nocturnal) and hence relatively small and inaccessible in high canopies; hunting them yields low energy returns. Indeed, Robert Bailey et al. (1989) found no unambiguous ethnographic accounts of foragers who lived in tropical rain forests without some reliance on domesti-cated plants and animals. In contrast, grasslands and open woodlands offer excel-lent opportunities for both collecting and hunting. Grasses, perennial herbaceous plants, shrubs, and dispersed trees store much less energy per unit area than a dense forest—but a higher share of it comes as easily collectible and highly nutri-tious seeds and fruits or in concentrated patches of large starchy roots and tubers.

Unlike animals in forests, many herbivores grazing on grasslands (including tundras) can grow to very large sizes. They often move in massive herds and can give excellent returns on energy invested in the hunt. Foraging hominids could secure meat on grasslands and woodlands even without any weapons, either as scavengers or as unmatched runners. Given the unimpressive physical endow-ment of early humans and the absence of effective weapons, it is most likely that our ancestors were initially much better scavengers than hunters (Blumenschine and Cavallo 1992). Large predators—lions, leopards, and saber-toothed cats—of-ten left behind partially eaten herbivore carcasses. This meat, or at least the nutri-tious bone marrow, could be reached by alert and enterprising early humans be-fore it was devoured by vultures and hyenas.

Humans can sweat more profusely than any other mammal. This feature, plus human bipedalism, made it possible for early hominids to chase even the fastest herbivores to exhaustion (see A2.1). David Carrier (1984) believes that this ability provided a notable evolutionary advantage that served our ancestors well in ap-propriating a new niche as diurnal, hot-temperature predators. The human abil-ity to work hard in hot environments was retained during migrations to colder climates. After a short acclimatization, northerners can match the sweating rates of hot-climate natives. This adaptation was apparent later in history: Without it, it would have been much harder to establish the post-1500 European empires.

The energy imperatives of hunting also made an incalculable contribution to human socialization. Because individual success in hunting large animals was fairly low, viable hunting groups had to maintain minimum cooperative sizes. The energies of more than a single family were also needed to track wounded animals, to butcher them, and to transport their meat. There was an obvious incentive to band together and then to pool the gains. Communal hunting received by far the greatest rewards. Working together, the hunters could execute well-planned strategies with considerable skill. For example, they could herd animals into confined runs (using brush and stone drive lines, wooden fences, and ramps) and capture them in prepared pens or natural traps or stampede them over cliffs (Frison 1987). Many large herbivores—mammoths, bison, deer, antelopes, and mountain sheep—could be slaughtered in these ways to provide caches of meat that could be frozen (in cold climates) or processed (smoked, pemmican).

All preagricultural societies were omnivorous. Although the members of such societies ate a large variety of plant and animal species, a few principal foodstuffs usually dominated their diets. The preference for seeds among gatherers was inevitable. Besides being rather easy to collect and store, seeds combine a high energy content with relatively high protein shares. Wild grass seeds have as much food energy as cultivated grains; nuts have energy densities up to 75 percent higher. All wild meat is an excellent source of protein, but most of it contains very little fat and hence has a very low energy density (less than half that of grains).

Not surprisingly, there has been a widespread hunting preference for large and relatively fatty species. A single small mammoth provided as much edible energy as fifty reindeer; a bison was easily equal to twenty deer (see Figure 2.2; A2.2). This is why our Neolithic ancestors were willing to ambush huge mammoths with their simple stone-tipped weapons and why Indians on North American plains, seeking fatty meat for pemmican, spent so much energy in pursuing bison.

But energy considerations alone cannot provide a full explanation of foraging behavior. If such concerns were always dominant, then optimal foraging—whereby gatherers and hunters tried to maximize their net energy gain by minimizing their investment of time and effort—would have been a universal strategy (Bettinger 1991). Optimal foraging explains the preference for large, fatty mammals or for less nutritious plant parts that do not need any processing (as opposed to energy-dense nuts, which may be hard to crack). Many foragers undoubtedly behaved in ways that generally maximized their overall rate of energy return, but other existential imperatives often worked against such behavior. Most notably, a society needs safe night shelters, a means of defending its territory against competing groups, reliable water sources, and adequate sources of vitamins and minerals. In addition, foraging behavior would be influenced by food preferences and attitudes toward work (see A2.3).

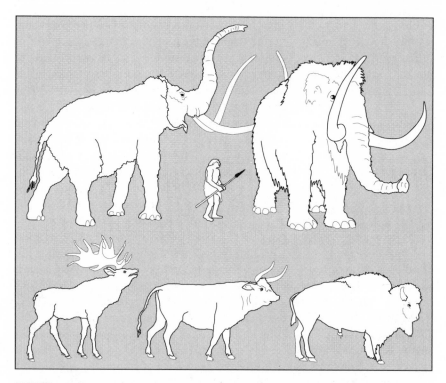

FIGURE 2.2 Scaled drawings of large prehistoric animals hunted in North America and Eurasia. Clockwise from top left: steppe mammoth (*Mammuthus columbi*), the most common giant mammal of prehistoric North America; woolly mammoth (*Mammuthus primigenius*), abundant on both continents during the last Ice Age; North American bison (*Bison bison*); auroch (*Bos primigenius*), the wild cattle of Europe and Asia; and giant European deer (*Megaloceros giganteus*). A hunter—1.6 meters tall—provides a comparison. *Sources:* Based on Steel and Harvey (1979) and Grzimek (1972).

It is impossible to reconstruct prehistoric energy balances, and this problem has provoked some scholars to make inadmissible generalizations. For some groups the total foraging effort was rather low, only a few hours per day. This fact, confirmed by some modern field surveys, led to the portrayal of foragers as "the original affluent society" living in a kind of material plenty filled with leisure and sleep (Sahlins 1972). This conclusion, based on very limited and dubious evidence, must be challenged (Birt-David 1992). Such theorizing ignores the reality of much of the hard and often dangerous work of foraging. More important, it ignores the frequency with which environmental stresses ravaged most foraging societies. Seasonal food shortages forced the members of these societies to eat unpalatable plant tissues and often led to weight loss or even devastating famines. They also resulted in high infant mortality rates (and also infanticide) and promoted low fertility (Johnson and Earle 1987).

Approximate calculations for a small number of twentieth-century foraging groups showed that the highest net energy returns could be obtained by gathering roots. As many as 30 to 40 units of food energy were acquired for every unit expended. In contrast, many hunting forays, above all those for smaller arboreal or ground mammals in the tropical rain forest, resulted in a net energy loss or bare equivalence (see A2.4). Typical gathering returns were ten- to twenty-fold, similar to those gained by hunting large mammals. The returns were undoubtedly much higher than this for many prehistoric societies dwelling in biomass-rich environments, and these higher returns would have allowed for gradual increases in social complexity. In fact, many foraging societies reached levels of complexity usually associated only with later agricultural societies. They had permanent settlements, high population densities, large-scale food storage, social stratification, elaborate rituals, and incipient crop cultivation.

The existence of such societies has been documented first in the Upper Paleolithic (Price and Brown 1985). Their sedentation and relative complexity were the result of unusually productive environments. Mammoth hunters in the Moravian loess region had well-built stone houses, produced a variety of excellent tools, and could fire clay (Klima 1954). The well-documented social complexity of the Upper Paleolithic groups of southwestern France was based on the strong Atlantic influence, which brought fairly cool summers but exceptionally mild winters.

This climate considerably extended the growing season and intensified the productivity of the continent's most southerly open tundra or steppe-like vegetation to support herbivorous herds larger than those anywhere in periglacial Europe (Mellars 1985). The complexity of these European Paleolithic cultures is best attested by their remarkable sculptures, carvings, and engravings. Such societies were slow in turning to agriculture. Farming was known in Europe's northern and eastern forest zone as early as between 5000 and 4500 B.C., but high foraging productivities in the West put off its adoption by two to three millennia (Zvelebil 1986).

Generally the highest productivities in complex foraging were associated with exploitation of fresh water or marine resources (Yesner 1980). A convincing illustration of this adaptation comes from excavations of the Mesolithic sites in southern Scandinavia (Price 1991). As the postglacial hunters depleted the stocks of large herbivores, they became hunters of porpoises and whales, fishermen, and collectors of shellfish. They lived in larger, more elaborate, and often permanent settlements. Many later Pleistocene and Holocene foraging societies were fairly large and sedentary, exhibited significant territoriality, and engaged in commercial exchange.

Similarly, fishing was the key to the largely settled way of life enjoyed by tribes in the Pacific Northwest. Their settlements commonly consisted of several hundred people living in well-built wooden houses. Regular runs made various salmon species a reliable and relatively easily exploitable resource. Salmon could be safely stored (when smoked) to provide an uncommonly good nutrition.

Thanks to its high fat content (about 15 percent) salmon has an energy density nearly three times higher than cod. But the Inuit of northwestern Alaska provide an even better example of a society with a high population density that is dependent on maritime hunting. The net energy returns of killing migrating baleen whales were more than 2000-fold (Sheehan 1985; see A2.5).

Societies dependent on a few seasonal energy flows for their food supply had to develop extensive and often elaborate methods of storage. These practices included caching in permafrost, drying and smoking (for seafood, berries, and various meats), storing (seeds and roots), preserving in oil, and making intestinal sausages, nut meal cakes, and flours (Hayden 1981). Large-scale, long-term food storage changed foragers' attitudes toward time, work, and nature and helped to stabilize populations at higher densities (Testart 1982). Such societies also learned how to plan and budget their time—perhaps the most important evolutionary benefit of the need for food storage.

Once established, storage-supported sedentation could not be abandoned without returning to lower population densities. This new mode of existence precluded frequent mobility and fostered the emergence of new activities and opportunities. Human existence shifted to a fundamentally different way of subsistence, and widespread surplus accumulation became the norm. The process was clearly self-amplifying: The human quest to manipulate an ever larger share of solar energy flows set the societies on the road toward higher complexity.

Origins of Agriculture

What we are dealing with is a complex, multifaceted adaptive system, and in human adaptive systems ... single all-efficient "causes" cannot exist.
—Bennet Bronson, in *Origins of Agriculture* (1977)

Why did foragers start to farm? Why did these new farming practices diffuse so widely, and why did their adoption proceed at what is, in evolutionary terms, a fairly rapid rate? I could sidestep these challenging questions by agreeing with David Rindos (1984) that "a process such as agriculture has no single cause, only a history with a multitude of interdependent interactions" (p. 769). Or I could select an appealing choice among the theories about the origin of agriculture surveyed by Charles Reed (1977) and Frederic Pryor (1983).

Overwhelming evidence for the evolutionary character of agricultural advances makes it possible to narrow down the possibilities. The most persuasive explanations of agricultural origins are those that give primacy to the combination of population growth and environmental stress (Wright 1971; Cohen 1977). Diminishing returns in gathering and hunting led to a gradual extension of the incipient cultivation present in many foraging societies. As a result, foraging and cultivation coexisted for very long periods. In Tell Abu Hureyra in northern Syria, an excavated Neolithic site revealed that hunting remained critical for the community's subsistence for 1000 years after the onset of plant domestication (Legge and Rowley-Conwy 1987).

No sensible explanation of agriculture's origins can ignore the many social advantages of farming. Sedentary crop cultivation provided an efficient way for more people to stay together and made it easier to have larger families, accumulate material possessions, and organize for both defense and offense. I would not go as far as Bryony Orme (1977) in concluding that "food production as an end in itself may have been unimportant" (p. 48). But there is no doubt that both the genesis and diffusion of agriculture had critical social cofactors, some of which were altogether independent of food provision.

Any explanation of agriculture that attributed its success primarily to higher energy returns would be too simplistic. The fact is that the net energy returns of early farming were often not at all superior or even equal to those of earlier or concurrent foraging activities. Compared to foraging, early farming usually required higher human energy inputs—but it could support higher population densities and provide a more reliable food supply. This explains why so many foraging societies had continuous interaction (and often much trade) with neighboring farming groups for thousands, or at least hundreds, of years before they adopted permanent farming practices of their own (Headland and Reid 1989).

There was no single center of plant and animal domestication. Food production appears to have started along the margins, rather than in core areas, of optimal zones (Wright 1971). In some nearly ideal regions—on the fertile floodplains of Mesopotamia, the Nile, the Indus, and the Huang He, for example, where seasonal floods annually replenish soil moisture and nutrients—the cultivation of wheat, barley, millet, and rice became permanent not long after its beginnings, that is, between 6000 and 9000 years ago.

Elsewhere in Asia, as well as in prehistoric Europe, Africa, and the Americas, early agriculture almost invariably took the form of shifting cultivation. This practice alternated short cropping periods of one to three years with fairly long fallow spells of a decade or more. In spite of many regional and local differences, many of the methods these societies employed were fundamentally the same (Clark and Haswell 1970). The cycle started with the clearing of natural vegetation, most often a forest. Slashing or burning the vegetation was often sufficient to prepare the surface for planting. Some plots had to be fenced to prevent damage by animals: In that case trees had to be felled for fences—a very labor-intensive task. Plant nitrogen was largely lost in combustion during the slash-and-burn process, but mineral nutrients enriched the soil and helped to produce good crops.

Grains (wheat, millet, rice, corn), roots (sweet potatoes, cassava, yams), and legumes (various beans, peanuts) dominated the cultivation. In warmer regions there was much interplanting, intercropping, and staggered harvesting. Reconstructions of net energy returns in traditional shifting farming show ranges of be-tween 11 and 15 for small grains (from Southeast Asian rices to African millets), 20 to 40 for most root crops, bananas, and good corn yields. Maxima close to 70 were obtained for roots and legumes in the best locations (see A2.6). Feeding one per-son required between 2 and 10 hectares of land, with the actually cultivated area ranging from just one-tenth of a hectare to 1 hectare. Even moderately productive shifting agriculture could support population densities an order of magnitude higher than those possible under the best-off sedentary foraging communities.

In areas of low population density and abundant land availability, shifting agri-culture was obviously part of the evolutionary sequence that went from foraging to incipient farming to permanent cropping. Shifting cultivation still feeds tens of millions of people in parts of Africa, Southeast Asia, and Latin America, but it is rapidly declining.

Where scarcity of precipitation, or its long seasonal absence, made cropping unrewarding or impossible, nomadic pastoralism was an effective alternative. An-imals can convert grasses into milk, meat, and blood with remarkably low inputs of human energy. Traditional nomadic pastoralists did not try to improve their grazing lands or employ selective breeding or disease control. Their labor was confined to herding the animals, guarding them against predators, watering them, helping with difficult deliveries, milking regularly and butchering infre-quently, and sometimes building temporary enclosures. The sustainable popula-tion densities of such societies were no higher than those of foraging groups (see A2.7).

For millennia nomadic grazing dominated parts of Europe and the Middle East and large regions of Africa and Asia. In all of these places it sometimes blended into mixtures of semi-nomadic agro-pastoralism, and in parts of Africa it coexisted with a significant degree of foraging. Often hemmed in by more pro-ductive farmers, and commonly dependent on barter with settled societies, some of these nomads had little impact beyond their confined worlds. But many groups exercised great influence on the Old World's history through their repeated inva-sions and temporary conquests of agricultural societies (Grousset 1970; Khazanov 1984). Pure pastoralists and agro-pastoralists survive even today—above all in Central Asia and in Sahelian and Eastern Africa—but they have an increasingly marginal existence.

APPENDIXES

A2.1 Running and Heat Dissipation by Humans

All quadrupeds have optimum speeds for different gaits (for example, horses can walk, trot, and canter). For common velocities of between 2 and 6 meters per second (m/s), the human costs of running are essentially independent of speed. Bipedalism and efficient heat dissipation explain this advantage. Ventilation in quadrupeds is limited to one breath per locomotor cycle: The thorax bones and muscles must absorb the impact on the front limbs as the dorso-ventral binding rhythmically compresses and expands the thorax space. In contrast, human breath frequency can vary relative to stride frequency. People thus can opt to run at a wide variety of speeds.

The extraordinary human ability to dispose of metabolic heat rests on very high rates of sweating. Horses can lose water at the rate of 100 grams per square meter (g/m^2) of their skin every hour, and camels lose up to 250 g/m^2, but people can perspire more than 500 g/m^2 (Hanna and Brown 1983). This perspiration rate translates into a heat loss of between 550 and 625 W, enough to regulate temperatures even during extremely hard work. People can also drink less than they perspire and make up for temporary partial dehydration within the next few hours. Consequently, the Tarahumara Indians of northern Mexico could run down deer; the Paiutes and Navajos could do the same with pronghorn antelopes. Kalahari Basarwa could chase duikers, gemsbok, and, during the dry season, even zebras, and some Australian Aborigines could chase down kangaroos. In hot weather, even excellent animal runners cannot match man's ability to run at variable speeds and dissipate heat.

A2.2 Body Mass, Energy Density, and Food Content of Animals[a]

Animals	Body Mass (kg)	Energy Density (MJ/kg)	Food Energy per Animal (MJ)
Whales	5000–40,000	25–30	80,000–800,000
Large proboscids (elephants, mammoths)	500–4000	10–12	2500–24,000
Large bovids (aurochs, bison)	200–400	10–12	1000–2400
Large cervids (elk, reindeer)	100–200	5–6	250–600
Seals	50–150	15–18	500–1800
Small bovids (deer, gazelles)	10–60	5–6	25–180
Large monkeys	3–10	5–6	5–30
Lagomorphs (hares, rabbits)	1–5	5–6	3–15

[a]I assume that the edible portion of whales and seals is about two-thirds of the animal and that the edible portion of other animals is about one-half the animal. The average energy density for whales is calculated by assuming that one-quarter of their body mass is blubber.

Sources: Based on data in Eisenberg (1981), Jarman et al. (1982), Sanders et al. (1979), and Sheehan (1985).

A2.3 Food Preferences and Attitudes About Work

Food preferences can be vastly different even among societies that share many other characteristics. This principle can be convincingly illustrated by a comparison between two highly similar foraging groups. The !Kung Basarwa owe their notoriety in anthropological literature to their dependence on the abundant and highly nutritious *mongongo* nut. This food source gave them better gross energy returns than any other food they could gather. But the /Aise, another Basarwa group with access to the same nuts, did not eat them because to them they did not taste good (Hitchcock and Ebert 1984).

An excellent example of how cultural realities can be at variance with simplistic energy models is provided by Jacques Lizot's (1977) comparison between two groups of the Yanomami Indians (northern Amazonia). One group was surrounded by a forest that was considerably richer in wild pigs, tapirs, and monkeys than the neighboring areas. The other group lived in an area less well endowed. The two groups possessed similar hunting skills and tools. Nevertheless, the first group consumed less than half of the amount that the neighboring group consumed in animal food energy and protein.

Lizot's explanation: The people of the first group were simply lazier. They hunted infrequently and for shorter periods and preferred to eat less well: "During one of the weeks … the men did not go hunting once, they had just collected their favorite hallucinogen (*Anadenanthera peregrina*) and spent entire days taking drugs; the women complained there was no meat, but the men turned a deaf ear" (p. 512). In this case, as in countless other instances in humanity's past, the variation in energy provided by hunting bore no relationship to the presence of animals or to the energy cost of the hunt but was simply a function of the attitude toward work.

A2.4 Net Energy Returns in Foraging

I use the method described in A1.10. Given the smaller prehistoric statures, I assume an average adult weight of just 50 kg. This weight would have required a basal metabolic rate of about 5.3 megajoules per day (MJ/d) and a minimum food intake of about 8 MJ/d, or roughly 330 kilojoules per hour (kJ/h). Plant collecting required mostly light labor, while hunting and fishing tasks ranged from light to heavy.

Typical foraging activities needed about four times the basal metabolic rate for men and five times the rate for women, or almost 900 kJ/h. Subtracting the basic existential need puts the net energy input of foraging at nearly 600 kJ/h. Energy output is simply the value of edible portions of collected plants or killed animals.

A2.5 Alaskan Whalers

Taking advantage of the annual migration of baleen whales, in fewer than four months of coastal hunting Alaskan hunters were able to amass enough food for settlements whose precontact population totaled about 2600 people (Sheehan 1985). The baleen whale has a huge mass—even an immature two-year-old averages nearly 12 tonnes. Its high food energy value—about 36 MJ/kg for blubber, 22 MJ/kg for *muktuk,* or skin and blubber—resulted in an enormous energy gain (more than 2000-fold). This food supported large permanent settlements and helped the culture attain an impressive degree of social complexity.

A2.6   Energy Costs and Population Densities in Shifting Cultivation[a]

Populations	Main Crops	Labor Inputs (hours)	Energy Returns	Population Densities (people/ha)
Southeast Asia	Tubers	2000–2500	15–20	0.6
Southeast Asia	Rice	2800–3200	15–20	0.5
West Africa	Millet	800–1200	10–20	0.3–0.4
Mesoamerica	Corn	600–1000	25–40	0.3–0.4
North America	Corn	600–800	25–30	0.2–0.3

[a]I assume an average labor input of 700 kJ/h. Outputs are edible harvests uncorrected for storage losses and seed needs.

Sources: Calculated from data in Allan (1965), Carter (1969), Clark and Haswell (1970), Conklin (1957), Heidenreich (1971), and Rappaport (1968).

A2.7 Nomadic Pastoralists

Johan Helland (1980) illustrates the low labor requirements of traditional nomadic pastoralism by pointing out the large numbers of major livestock species managed by a single herder in the East African setting: up to 100 camels, 200 cattle, and 400 sheep and goats. Anatoli Khazanov (1984) lists similarly large figures for Asian pastoralists: two mounted shepherds for 2000 sheep in Mongolia, an adult shepherd and a boy to tend 400 to 800 cattle in Turkmenia. The appeal of low labor needs was one of the key reasons for pastoralists' reluctance to convert to farming. But nomadic population densities were also low, typically just between 1.1 and 2.2 people per square kilometer (km^2).

3

Traditional Agriculture

Tillage is breaking and dividing the Ground by Spade, Plow, Hoe, or other Instruments ...

By Dung we are limited to the Quantity of it we can procure, which in most places is too scanty.

But by Tillage, we can enlarge our Field of Subterranean Pasture without Limitation.

—Jethro Tull, *The Horse Hoeing Husbandry* (1733)

IN SPITE OF MANY differences in agronomic practices and in cultivated crops, all traditional agricultures shared the same energetic foundations. They were powered by the photosynthetic conversion of solar radiation. Photosynthesis produced food for people, feed for animals, recycled wastes for the replenishment of soil fertility, and fuels for smelting the metals needed to make simple farm tools. Consequently, traditional farming was fully renewable. With the exception of cutting down old forests, there was no depletion of accumulated energy stocks. The whole enterprise relied on virtually immediate conversions of solar energy flows.

And yet this renewability was no guarantee of sustainability. Poor agronomic practices could lower soil fertility or cause excessive erosion or desertification. Such environmental degradations could lower yields or even cause the abandonment of cultivation. But in most regions, traditional farming progressed from extensive to intensive cultivation. Its prime movers—human and animal muscles—remained unchanged for millennia, but cropping practices, cultivated varieties, and the organization of labor were subject to great transformations. Thus, both constancy and change marked the history of traditional farming.

The advancing intensification of farming—that is, increases in fertilization and irrigation and more frequent cropping and plant rotations—sustained higher population densities but demanded higher energy expenditures. This principle applied not only to direct farming activities but also to such critical supportive measures as the digging of wells and the building of irrigation canals, terraced fields, roads, and food storage facilities. In turn, these improvements required more energy for producing a larger variety of better tools and simple machines powered by domestic animals or by water and wind.

The more intensive forms of cultivation relied on animal labor at least for plowing, usually by far the most energy-demanding task in field farming. The intensive agricultures of the Americas were notable exceptions: Neither Mesoamerican cultivators nor Inca potato and corn growers had draft animals. Feeding large numbers of domestic animals required further intensification to produce feed crops. In all better-off traditional agricultures, animals were also used extensively for many other field tasks as well as in grain threshing and milling. They were, of course, indispensable for the land-borne distribution of food. Stabling, feeding, and breeding these animals and producing harnesses, shoes, and implements for them introduced new complexity and skills to agricultural societies.

Intensified farming also required preindustrial civilizations to conduct long-range planning, make investments for the future, and establish new forms of labor organization. Not every form of cropping intensification required centralized organization and oversight. Some practices, such as digging short, shallow irrigation canals or wells or building a few terraces or raised fields, originated repeatedly with individual peasant families or villages. But eventually the scale of irrigation, drainage, terracing, or raised field construction demanded hierarchical coordination and supra-local management. Such organized efforts also went into projects like building granaries, roads, and shipping canals. And eventually, more powerful energy sources were needed to process larger amounts of grain and oilseeds for growing cities. This demand was an important stimulus for the development of the first important substitutes for human and animal muscles—the use of water and wind flows for milling grain and pressing oilseeds.

In this chapter I will examine traditional farming in four old high cultures—Egypt, China, Mesoamerica, and Europe—as well as in the United States in the nineteenth century. I will look at particular agronomic practices and cropping systems in these cultures and calculate energy budgets of principal cropping practices. These traditional agricultures shared many commonalities, but their differences are noteworthy. The productive limits of all of them were removed only by the rising inputs of fossil fuels.

Commonalities and Peculiarities

It is a certain thing that the chiefest and fundamentallest point in husbandry is to understand the nature and condition of the land that one would till, and to sow it with such seeds as it will produce either naturally or by art that which may turn to a man's greatest profit and advantage.

—Sir Richard Weston, *A Discours of Husbandrie*
used in Brabant and Flanders (1652)

The requirements of crop growing impose a general pattern on the sequence of field work. Moreover, cultures that cultivated identical crops invented or adopted very similar agronomic practices, tools, and simple machines. Some of these innovations came early, diffused rapidly, and then remained largely unchanged for millennia. Others were unique to their original regions for a very long time but, once diffused, underwent rapid improvement. The sickle and the flail are in the first category; the iron moldboard plow and the seed drill are in the second.

Before discussing the individual cultures, I will first provide brief but systematic descriptions of all important common field operations, tools, and simple machines. I will then turn to the most important commonality, the dominance of cereal grains in traditional farming, and to peculiarities of cropping cycles.

Field Work

See that you keep your ploughs and ploughshares in good condition. ... See that you carry out all farm operations betimes, for this is the way with farming: if you are late in doing one thing you will be late in doing everything.

—Marcus Cato, *De agri cultura* (second century b.c.)

In all of the Old World's high cultures, crop cultivation started with plowing. Its indispensability is reflected even in the oldest writing. Both the Sumerian cuneiform records and the Egyptian glyphs have pictograms for plows (Jensen 1969). Plowing prepares the ground for seeding much more thoroughly than hoeing does: It breaks up the compacted soil, uproots established plants, and provides weed-free, loosened, well-aerated ground in which seedlings can germinate and thrive.

The first primitive scratch plows (ards), commonly used shortly after 4000 b.c. in Mesopotamia, were pointed wooden sticks with a handle. Later most of them were tipped with metal. For centuries they remained lightweight and symmetrical (with the draft line in a vertical plane with the beam and share point). (See Figure 3.1.) Such simple plows, which merely opened up a shallow furrow for seeds and left cut weeds on the surface, were the mainstay of both Greek and Roman farming. They were used over large parts of the Middle East, Africa, and Asia until the twentieth century. In the poorest regions they were pulled by people. Only in lighter, sandier soils would such an effort be speedier than hoeing (Bray 1984).

The addition of a moldboard was by far the most important improvement. A moldboard guides the plowed-up soil to one side, turns it (partially or totally), buries the cut weeds, and cleans the furrow bottom for the next turn. A moldboard also makes it possible to till a field in one operation, whereas ards require cross-plowing. With a moldboard, the plow is asymmetrical because its draft line is displaced slightly toward the side of the turned-up soil. The first moldboards were straight pieces of wood, but before the first century b.c. the Han Chinese introduced curved metal plates joined to the plowshare (see Figure 3.2). This fundamental innovation spread beyond East Asia only some seventeen centuries later.

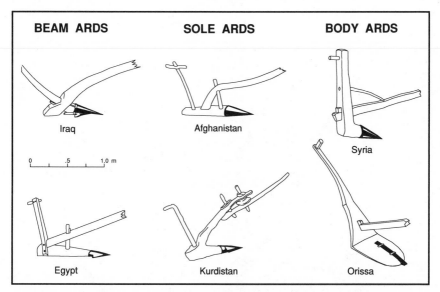

BEAM ARDS   **SOLE ARDS**   **BODY ARDS**

Iraq   Afghanistan   Syria

Egypt   Kurdistan   Orissa

FIGURE 3.1 A variety of ards. Traditional farmers used these light, wooden, symmetrical plows for working loose soils. Beam ards were the earliest types. In sole ards the beam and handle were separately inserted into the sole body, whereas body ards tapered off into the handle. *Source:* Adapted from Hopfen (1969).

After the introduction of Chinese curved moldboards to Europe (by Dutch sailors in the seventeenth century), European and American designs improved rapidly. During the second half of the eighteenth century Western plows still retained their heavy wooden wheels, but they carried well-curved iron moldboards. These moldboard plows became especially efficient when steel replaced cast iron beginning in the 1830s. In most soils plowing leaves behind relatively large clods, which must be broken before seeding. Hoeing will work, but it is too slow and too laborious. That is why various forms of harrows have been used by all old plow cultures (see Figure 3.3). Inverted harrows or rollers were often used to further smooth the planting surface.

After plowing, harrowing, and leveling, the ground was ready to be seeded. Although seed drills were used in Mesopotamia as early as 1300 B.C., and sowing plows were used by the Han Chinese, broadcast seeding by hand—a wasteful method that resulted in uneven germination—remained common in Europe until the nineteenth century. Simple drills were eventually developed that dropped seeds through a tube from a bin attached to a plow. These drills spread from northern Italy starting in the late sixteenth century. Many innovations turned them before too long into complex seeding machines. The intercultivation of growing crops—loosening the rain-compacted soil, destroying weeds, and forming ridges—was done largely by hoeing. Peasants brought manures and other organic wastes to the fields on carts, in wooden cisterns, or in buckets carried at the

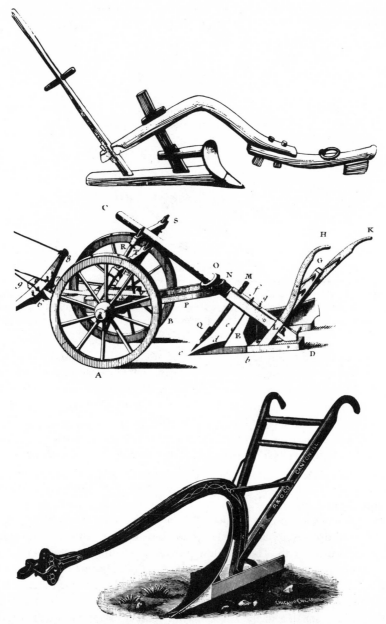

FIGURE 3.2 Evolution of curved moldboard plows. In traditional Chinese plows (*top*), relatively small but smoothly curving moldboards were made from nonbrittle cast iron. Early European adaptations, attached to a heavy medieval forecarriage (*bottom left*), had a pointed coulter placed in front of the share to cut the roots. The efficient American steel beam plow of the mid-nineteenth century (*bottom right*) had its share and moldboard fused in a smoothly curving steel triangle. *Sources:* Adapted from Hopfen (1969), Diderot and D'Alembert (1769–1772), and Ardrey (1894).

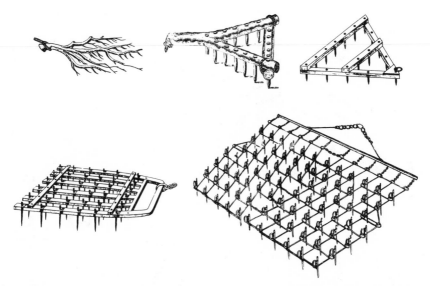

FIGURE 3.3 Different types of harrows, including (*top row, left to right*) a forked branch harrow, a simple Y-shaped Roman spike-toothed harrow, and a common nineteenth-century wooden, steel-spiked harrow. These simple harrows led to a variety of metal frames to which were fastened metal teeth or discs (*bottom*). *Source:* Reproduced from Ardrey (1894).

end of shoulder beams. Then they used pitchforks or simply poured or ladled the material onto the field.

Sickles were the first harvesting tools to replace the short, sharp stone cutters used by many foraging societies. Serrated (in the oldest designs) or with smooth edges, with semicircular, straight, or slightly curved blades, they have been used in countless variations ever since. Cutting with sickles was very slow; traditional farmers preferred the much more efficient scythes, equipped with cradles for grain reaping, for harvesting larger areas (see Figure 3.4). But slow sickle harvesting caused fewer grain losses due to ear shattering than the faster scythe cutting did. This is why it was retained throughout Asia in harvesting easily shattered rice. Mechanical reapers came only in the early part of the nineteenth century. At harvest, the sheaves were carried on the workers' heads, loaded into panniers and attached to shoulderbeams or carried by animals, or transported in wheelbarrows, carts, or four-wheeled wagons that could be pushed or pulled by people or draft animals.

Farmers expended a considerable amount of energy on crop processing. They spread the grain on a threshing floor and beat it with sticks or flails, and they struck sheaves against grates or pulled them across special combs. They trained animals to tread on the spread grain or to pull heavy sleds or rollers over it. Before the adoption of crank-turned fans, farmers manually winnowed chaff and dirt

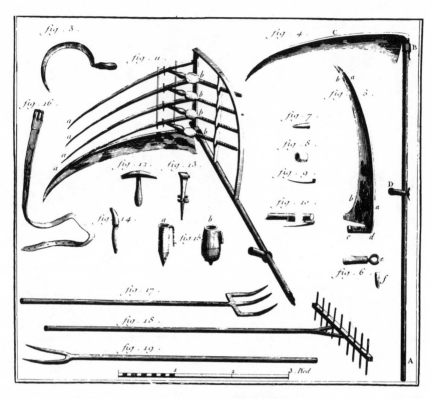

FIGURE 3.4 Sickles and scythes. The simple scythe on the right was used for grass mowing, the cradled one for cereal harvesting. Also pictured are tools for hammering and sharpening the scythes, as well as a rake and pitchforks. The bottom illustrations show the nineteenth-century American harvesting of grain by sickle and cradled scythe. *Sources:* Top illustration reproduced from Diderot and D'Alembert (1769–1772); bottom illustrations from materials of the Massey-Harris Company.

from the grain using baskets and sieves. Tedious manual labor also went into milling the grain before animals, water, and windmills mechanized the task. Manual or animal-operated presses were used to extract the oil from seeds and the sweet juice from cane.

Dominance of Grains

If there's a mountain, we'll cover it with wheat.
If there's water to be found, we'll use it all to plant rice.
—Lu Yu, *The Farmer's Lament* (twelfth century)

Although all traditional agricultures grew a variety of grain, oil, fiber, and feed crops, the sequence of common field tasks described above was performed most often in cultivating cereals. Cereal grains dominated annual cropping in all Old World agricultures and were even more widespread than plowing. Plowless Mesoamerican societies relied on corn, and even the Incas, who planted a wide variety of potatoes at high elevations and on steep mountain slopes, cultivated corn in lower altitudes and grew quinoa on the high-lying Andean Altiplano. There were many cereals that were only locally or regionally important, but the main genera have gradually diffused worldwide from their areas of origin. Wheat spread from the Near East, rice from Southeast Asia, corn from Mesoamerica, and millet from China (Vavilov 1951).

A combination of evolutionary adjustments and energetic imperatives is responsible for the importance of grain in the human diet. Foraging societies depended on a wide variety of plants, and tubers or seeds provided most of their food energy. In settled societies, the tuber option was highly restricted. The water content of freshly harvested tubers is too high for long-term storage in the absence of effective temperature and humidity controls. Even if this challenge were to be overcome, their bulkiness would require enormous storage volumes over the winter months, especially in the densely settled populations of the more northerly latitudes (or in higher altitudes). Incas solved the problem by preserving potatoes as *chuñu*. This dehydrated foodstuff, produced by trampling and by an alternating process of freezing and drying, is storable for years.

Without elaborate processing tubers could be relied upon only in the tropics, where their year-round harvest obviates storage. Even there they are a poor nutritional choice because of their low protein content (typically about one-fifth of the protein content of cereals). Leguminous grains (peas, beans, and lentils) have commonly twice as much protein as cereal grains (and soybeans have three times as much). The average wheat harvest yields much more food energy than the average harvest of legumes, however, and good corn and rice harvests are even more productive.

Cereal grains thus provide clear energy advantages. They offer fairly high yields, good nutritional value (high in filling carbohydrates, moderately rich in proteins), a relatively high energy density at maturity (roughly five times higher than for tubers), and a low moisture content, which makes them suitable for long-term storage. The regional dominance of a particular species is largely a matter of environmental limits and taste preferences.

In terms of total energy content, all cereals are remarkably similar. The bulk of the food energy is in carbohydrates, which are present mostly as polysaccharides (starches) and are highly digestible: People eating balanced diets can use virtually all of this energy in their metabolism.

The protein content of cereals ranges from less than 10 percent for many rices to almost 15 percent for hard wheat and up to 16 percent for quinoa. Proteins contain about one-third more energy than carbohydrates, but their primary role in human nutrition is not as a source of energy but of amino acids, which are essential to building and repairing body tissues. We cannot synthesize proteins without consuming these essential amino acids in the correct ratios. Although animal foods and mushrooms supply complete proteins, plant foods do not.

Even the strictest vegetarian can obtain complete protein, however, by combining foods with particular amino acid deficiencies. Cereal grains are always short in lysine—but legumes, short in methionine, have plenty of lysine. All traditional agricultural societies subsisting on largely vegetarian diets dominated by cereal grains found the solution in eating combinations of the two kinds of seeds. In China, soybeans, beans, peas, and peanuts traditionally supplemented millet, wheat, and rice. In India, protein from lentils, peas, and chick-peas enriched wheat and rice. In Europe, the combination was peas and beans with wheats, barley, oats, and rye; in West Africa, peanuts and cowpeas with millet; and in the New World, corn and beans were commonly intercropped, that is, grown together in alternating rows in the same field, and eaten together in a variety of dishes.

Two of the major proteins in wheat flour are unique, not nutritionally, but because of their physical properties. When combined with water they form a gluten complex that is sufficiently elastic to permit a leavened dough to rise—and yet strong enough to retain the carbon dioxide bubbles formed during the process of yeast fermentation (Ayres et al. 1980). Without wheat gluten, there would be no leavened bread, the basic food of Western civilization.

The energy balances of grain production are the most revealing indicators of agricultural productivity in traditional cultures. Data on typical agricultural labor requirements and their energy costs are available for a large variety of individual field and farmyard tasks (see A3.1). But this level of detail is not necessary for calculations of approximate energy balances. Using a representative average of typical net energy costs in traditional farming works quite well (see A3.2). So does the use of gross grain output, calculated by multiplying the mass of the harvested grain by appropriate energy equivalents. The ratio of the food energy in grain to the food energy for producing that grain indicates the gross energy return, and hence the productivity, of these critical farming tasks.

The net energy returns—after subtracting seed requirements and the inevitable milling and storage losses—were substantially lower. Farmers had to set aside a portion of every harvest for the next year's seeding. The combination of low yields and high seed waste in broadcasting could mean that as much as one-third or even one-half of medieval grain crops had to be set aside. Gradually, these shares fell to less than 15 percent. Before actual food preparation (cooking or baking), most cereals are milled. Whole grain flours incorporate the complete cereal, but most flours contain between 80 and 85 percent of the original grain. Rice loses even more of its mass during milling, usually between 25 and 35 percent.

Storage losses can commonly reduce the edible grain total by up to 10 percent. Grain with a moisture content of less than 13 percent can be stored for long periods of time, but higher moisture contents, especially when combined with higher temperatures, provide perfect conditions for seed germination, insects, and fungi. In addition, improperly stored grain can be consumed by rodents. Even during the eighteenth century a combination of seeding requirements and storage losses could have reduced the gross energy gain by around 25 percent.

Cropping Cycles

In the days of the third month we take our plows in hand …
In the fourth month the grass is in seed.
In the fifth month the cicada gives its note.
In the eighth, we reap,
In the tenth, the leaves fall.

—Shi jing (*Book of Poetry*) (eleventh century b.c.)

Despite the commonalities of cereal cultivation, annual crop cycles harbored an astonishing variety of local and regional peculiarities. Some of them had distinctly cultural origins, but most were responses and adaptations to environmental factors. Most notably, environmental conditions determined the choice of leading crops and hence the makeup of typical diets. They also molded the rhythm of annual farming cycles, which in turn determined the management of agricultural labor. Wheat, for example, spread from the arid Middle East to all the continents because it can do well in many climates (in semi-deserts as well as in rainy temperate zones), in many elevations (from sea level up to 3000 meters), and on many soils, as long as they are well-drained (Briggle 1980). Only the low-lying tropics are excluded because wheat is a cool season crop; for that very reason it became the leading food cultivar in the temperate zone between 30 and 60° N.

In contrast, rice is originally a semi-aquatic plant of the tropical lowlands. It grows in fields flooded with water until just before the harvest. Its cultivation has also spread far beyond the original South Asian core, but the best yields have always been in rainy tropical and subtropical regions. There are much higher labor requirements in rice cultivation than in wheat cropping because the former requires the construction and maintenance of ridged wet fields, the germination of seeds in nurseries, the transplantation of seedlings, and subsidiary irrigation. Like rice, corn yields its best harvests in regions with warm and rainy growing seasons, but it prefers well-drained soils. Potatoes grow best where summers are cool and rains abundant.

Annual farming cycles were governed by water availability in both the arid subtropics and monsoonal regions. In temperate climates, they were governed by the length of the growing season. Specific regions had unique requirements. For example, in Egypt the Nile's floods determined the annual cycle of cultivation. Sowing in moist soils started as soon as the water receded (usually in November), and no field work could be done between the end of June, when the waters started to rise, and the end of October, when they rapidly receded (Hassan 1984). This pattern prevailed largely unchanged until the adoption of widespread perennial irrigation in the nineteenth century.

In monsoonal Asia, rice cultivators had to use the abundant summer precipitation. For example, in intensive Chinese cropping, early rice seedlings were transplanted from nurseries to fields as early as April. After the first crop's harvest in July, farmers immediately transplanted the late rice seedlings. This crop would be harvested in the late fall and it was, in turn, followed by a winter crop. Double-cropping in temperate zones worked under much less pressure. In Western Europe the over-wintering crops planted in the fall were harvested five to seven months later. They were followed by spring-seeded crops, which matured in four to five months. In all cold northern regions, farmers could plant only a single annual crop, and they had to wait until the ground thawed and the danger of killing frosts receded. Consequently, the crop had only about three months in which to mature.

The rhythms of cultivation put highly fluctuating demands on the mobilization and management of human and animal labor. Regions with a single annual crop had long months of winter idleness. This was characteristic of grain farming in northern Europe and on the North American plains, for example. Taking care of domestic animals was of course a year-round task, but it still left much free time. Families spent this time doing domestic craft work, making repairs on farming equipment, building fences, and so on. In North China, many winter days were devoted to maintaining and extending irrigation works.

Spring plowing and seeding called for a few weeks of hard work followed by a few months of an easier routine. Harvest was the most taxing time, but the fall plowing could extend over a much longer period. Where less extreme climates allowed the planting of a winter crop—in Western Europe, on the North China Plain, and in most of eastern North America, for example—there were two to three months between harvesting the summer crop and putting in the winter one.

In contrast, in countries with much less evenly distributed precipitation, and especially in monsoonal regions, there were only limited slots of time available for performing field tasks. Because even short delays could be costly, timeliness in planting and harvesting was especially critical. Even a week's delay beyond the optimum planting period could cause a substantial reduction in yield. The early grain harvest sometimes needed to undergo a labor-intensive drying process because of its high moisture content. When a crop had to be harvested quickly so that the next one could be planted, the labor demand soared. An old Chinese proverb captured this need: "When both the millet and the wheat are yellow ripe, even the spinning girls have to come out to help."

Throughout the double-cropping area of China, between March and September traditional farming utilized on the average between 94 and 98 percent of the available labor supply (Buck 1937). In parts of India that have a very short rainy season, the two peak summer months required more than 110 or even 120 percent of actually available labor, and a similar situation existed in other parts of monsoonal Asia (Clark and Haswell 1970). This need could be met only if all members of a farming family worked arduously long hours—or by relying on migratory labor. These peak labor demands were among the most important energetic bottlenecks of traditional farming. Before the introduction of reapers and binders, manual grain harvesting was the most time-consuming task in traditional farming. The work took three to four times longer than plowing, and it put clear limits on the maximum area that a single small family could manage.

The use of animal labor, reserved in many agricultures for the most demanding field tasks, varied even more. For example, the maximum work periods for South China's water buffaloes included two months of planting, harrowing, and grading in the early spring, six weeks of harvesting in the summer, and a month of field preparation (again, plowing and harrowing) for a winter crop—altogether about 130 to 140 days. In the single-cropping regimes of northern Europe, draft horses would do only sixty to eighty days of strenuous field work; some of this time was for plowing and harvesting, but most of it was for transport. A typical working day ranged from just five hours for oxen in many African locations to more than ten hours for water buffaloes in Asian rice fields or for horses during European or North American grain harvests.

Routes to Intensification

Whatever grows will rise in mad confusion
And toil must guide the crop to its conclusion.

—Du Fu, *Directing Farmers* (eighth century)

No quest for higher yields could succeed without three essential advances. The first was a partial replacement of human work by animal labor. In rice farming, where deep plowing supplanted hoeing, this substitution eliminated only the most exhaustive types of human work. In dryland farming similar substitutions sped up many field and farmyard tasks, freeing people to pursue other activities. This prime mover shift did more than make the work quicker and easier: It also improved the quality of the work for tasks as diverse as plowing, seeding, and threshing.

Second, irrigation and fertilization moderated, if not altogether removed, the two key constraints on crop productivity, shortages of water and lack of nutrients. Third, growing a wider variety of crops, either by multicropping or in rotations, made traditional cultivation both more resilient and more productive. Two old Chinese peasant sayings convey the importance of these principles: "Whether there is a harvest depends on water; how big it is depends on fertilizer," and "Plant millet after millet and you will end up weeping."

The use of draft animals introduced fundamental energetic advances with implications beyond field cultivation and harvesting. Draft animals became indispensable for fertilization, both as sources of valuable plant nutrients in manures and as prime movers distributing them to crops. In many places draft animals were also the leading energizer of irrigation. More powerful prime movers and better water and nutrient supplies also expanded the possibilities of multicropping and crop rotations. In turn, these advances made it possible to support larger numbers of more powerful animals. Clearly, the three paths to intensification were linked by mutually reinforcing feedbacks.

Draft Animals

> *The first consideration, then, in the matter of quadrupeds, is the proper kind of ox to be purchased for ploughing. You should purchase them unbroken, not less than three years old and not more than four; they should be powerful and equally matched so that the stronger will not exhaust the weaker when they work together.*
>
> —Marcus Terrentius Varro,
> *De re rustica* (first century b.c.)

Thousands of years of animal husbandry have resulted in a large number of working breeds with distinctive characteristics. Size and weight varied widely. Whereas most Indian bullocks always weighed less than 400 kilograms, Italian Romagnola or Chianina draft cattle were easily twice as heavy (Rouse 1970). Most of the draft horses used in Asia and in many regions of Europe were, according to modern division, ponies: They weighed no more than Asian oxen and were less than 14 hands tall. In contrast, large adults of the heaviest European breeds (Belgian Brabançons, French Boulonnais and Percherons, Scottish Clydesdales, English Suffolks and Shires, German Rheinlanders) approached and even topped 17 hands and weighed around 1 tonne (Silver 1976). And water buffaloes could range from just 250 to 700 kilograms (Cockrill 1974).

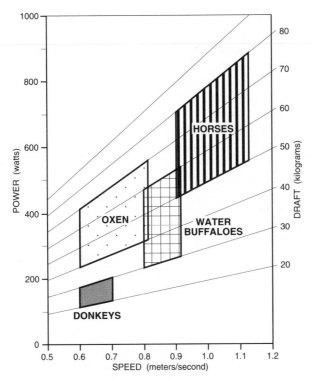

FIGURE 3.5 Comparisons of animal draft power. The data show the clear superiority of horses. *Sources:* Plotted from data in Hopfen (1969), Rouse (1970), and Cockrill (1974).

Plowing was undoubtedly the activity where animals made the greatest difference (Leser 1931). Given the relatively high power requirements of this task, it is hardly surprising that the first clearly documented cases of cattle domestication involved plow farming. In time, these animals were also used in many regions for lifting irrigation water and for processing harvested crops, and they were eventually used everywhere for transportation.

In general, the tractive force of working animals is roughly proportional to their weight. The other variables determining the actual performance include the animal's sex, age, health, and experience, as well as the efficiency of the harness and soil and terrain conditions. As all of these variables can vary rather widely, it is preferable to summarize the useful power of common working species in terms of typical ranges (Hopfen 1969; Rouse 1970; Cockrill 1974). The power of small donkeys was hardly above that of strong adult men, whereas oxen were commonly at least three times as powerful. Horses are superior draft animals, although most of them could never sustain prolonged output at or above 1 horsepower (see Figure 3.5; A3.3).

This superiority was not translated into actual performance until the invention and general adoption of an efficient harness (Haudricourt and Delamarre 1955; des Noettes 1931; Needham et al. 1965). Traction must be transferred to the point of work—be it plowshare or reaper's edge—by a gear that allows for efficient transmission of power and enables human control of the animal's movements. Such designs may look simple, but they took a long time to emerge. Cattle, the first working animals, were harnessed by various yokes. These were simple straight or curved wooden bars fastened to the animals' horns or necks (see Figure 3.6). The most ancient harness, of Mesopotamian origin and later common in Spain and Latin America, was the double head yoke, which was fixed either at the front or the back of the head. Best used with strong, short-necked animals, it was particularly inefficient with long-necked cattle.

A more comfortable single head yoke was used in several parts of Europe (eastern Baltic region, southwestern Germany). The single neck yoke, connected to two shafts or to traces and a swingletree, was common throughout East Asia and Central Europe. Africa, the Middle East, and South Asia favored a double neck yoke. This was a primitive harness: merely a long wooden beam with throat fastenings that could choke the animal in heavy labor and with a traction angle that was too large. Moreover, in order to avoid excessive choking, the oxen or cows had to be of identical height and a pair had to be harnessed even for lighter work that a single animal could do.

Horses are the most powerful draft animals. Unlike cattle, whose body mass is almost equally divided between the front and the rear, horses' fronts are notably heavier than their rears (by a ratio of about 3:2) and so the pulling animal can take better advantage of inertial motion (Smythe 1967). Except in heavy, wet soils, horses can work steadily at speeds of around 1 meter per second, easily 30 to 50 percent faster than oxen. Maximum two-hour pulls for paired heavy horses can be almost twice as high as for the best cattle pairs. The largest horses can thus work for short spells at rates surpassing 2 kilowatts, more than 3 standard horsepower!

The oldest existing images of working horses do not show them laboring in fields but rather pulling light ceremonial or attack carriages (see Figure 3.7). The throat-and-girth harness used for such purposes would not do for heavy field tasks. It had two key drawbacks. One was that it had an excessively high point of traction. The other was that, in order to prevent the backward slippage of the girth, the throat-strap created a choking effect. As soon as the horse leaned into a heavy pull by lowering and advancing its head, the strap would constrict its breathing. The breastband harness, introduced in China sometime before the early Han dynasty, increased the efficiency of draft horses; however, its point of traction, although removed from the throat, was set too far away from the animal's powerful pectoral muscles. The design spread across Eurasia and reached Italy as early as the fifth century, most likely with migrating Ostrogoths; it reached northern Europe some 300 years later.

FIGURE 3.6 Head and neck yokes for working oxen. The head yoke (*top*), the first harness for working oxen, was highly inefficient. The neck yoke (*bottom*) became the dominant harness throughout the Old World. *Sources:* Adapted from Hopfen (1969) and from a late Ming dynasty (1637) painting.

FIGURE 3.7 Throat-and-girth and breastband harnesses. The throat-and-girth harness (*top*), here on two horses pulling a Greek chariot, was unsuitable for field work. The breastband harness (*bottom*) was superior to it and remained in use for lighter duty until the nineteenth century. *Sources:* Adapted from Jope (1956) and from Diderot and D'Alembert (1769–1772).

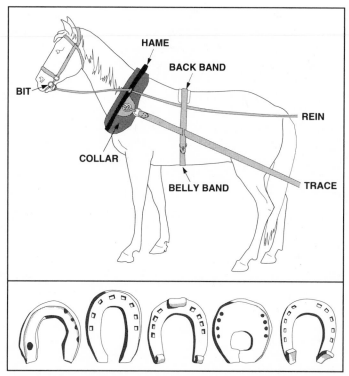

FIGURE 3.8 The collar harness and horseshoes transformed horses into heavy draft animals. The harness shows a typical nineteenth-century arrangement. A variety of mid-eighteenth-century horseshoes is shown at the bottom. The shapes are, respectively, the typical English, Spanish, German, Turkish, and French horseshoes. *Sources:* Harness based on Telleen (1977) and Villiers (1976); horseshoes reproduced from Diderot and D'Alembert (1769–1772).

It took another Chinese invention to turn horses into superior working animals. The collar harness was first used in China perhaps as early as the first century B.C. as a soft support for the hard yoke and only gradually was transformed into a single component. By the fifth century its simple variant could be seen on the Donghuang frescoes. Philological evidence suggests that by the ninth century it had reached Europe, where it was in general use within about three centuries; there, it remained largely unchanged until horses were replaced by machines more than 700 years later. Millions of China's working horses use it still.

A standard collar harness consisted of a single oval wooden frame (later also metal); it was lined for a comfortable fit on a horse's shoulders and often had a separate collar pad underneath. Draft traces were connected to the hame just above the horse's shoulder blades (see Figure 3.8). The horse's movements were controlled by the bridle: a metal bit in its mouth attached to reins and braced by a headstall. The collar harness provided a desirably low traction angle and allowed for heavy exertion. The horse could deploy its powerful breast and shoulder muscles without any restrictions on its breathing. Experimental comparisons have shown that the useful draft of horses harnessed with collars is at least four times, and in brief exertions up to ten times, higher than for animals in a throat-and-girth harness. The collar harness also allowed for effective teaming-up of animals in a single or double file for exceptionally heavy labor.

An efficient harness was not all that was needed to draw out the horse's superior performance. Horseshoes, narrow U-shaped metal plates fitting the hoof's rim and fastened with nails driven into the insensitive hoof wall, were also needed (see Figure 3.8). Their use prevented excessive wear of the animal's soft hooves and improved its traction and endurance. This invention was especially important in the cool, wet climate of western and northern Europe. The Greeks did not have metal horseshoes: They encased hooves in leather sandals filled with straw. Horseshoes became common in Europe only after the ninth century. Whipple-trees (or swingletrees) also enhanced horses' usefulness. Attached to traces, interconnected, and then fastened to field implements, they equalized the strain resulting from uneven pulling, made it easier to lead the animals, and allowed for harnessing an even or odd number of horses (see Figure 3.9).

Although power comparisons provide a clear ranking, there is no perfect all-purpose draft animal. The traditional reliance on various species and various types of harnesses resulted from an interplay of environmental and socioeconomic conditions. Besides greater power, horses also have better endurance and live longer than other draft animals. Workdays of eight to ten hours used to be the norm for horses, compared to four to six hours for cattle. Both oxen and horses started working at three or four years of age, but oxen usually lasted for just eight to ten years, whereas horses could commonly carry on for fifteen to twenty years.

Finally, the anatomy of a horse's leg virtually eliminates the energy costs of standing. This unique advantage is due to the very powerful suspensory ligament, which runs down the back of the cannon bone, and a pair of tendons (superficial and deep digital flexors) that can "lock" the limb without engaging muscles. This feature allows the animals to rest, even to doze, while standing with hardly any metabolic cost (Smythe 1967). Horses can thus spend little energy while grazing or resting in harness. All other mammals need about 10 percent more energy when standing as compared to lying down.

Powerful horses could be expensive to maintain. They had to be fed with grain, harnessed with special equipment, and shod. In contrast, working oxen, which were weaker and slower, could be fed with just straw and chaff and harnessed very cheaply. Moreover, horses do not thrive in all climates. Few horses do well in the tropics, whereas humpback cattle work well in high temperatures thanks to their efficient heat regulation; they are also less susceptible to tick infestation. And water buffaloes thrive in warm, wet environments that horses cannot tolerate. Water

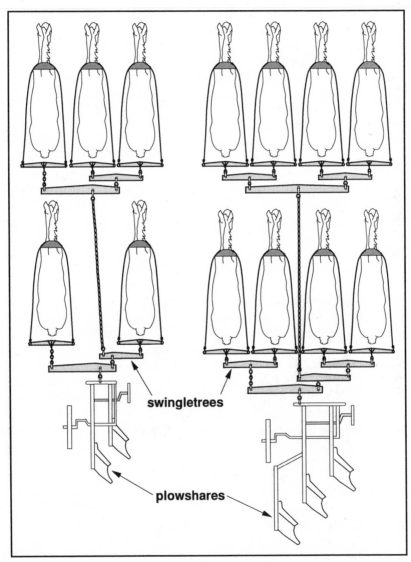

FIGURE 3.9 Whippletrees (swingletrees) made it possible to transmit draft power more efficiently by equalizing pulls and to harness any even or odd number of animals. *Source:* Based on Telleen (1977).

buffaloes also convert poor roughage feed more efficiently than cattle and can graze on aquatic plants while fully submerged.

Smaller animals also have some mechanical advantages. They deliver relatively more useful power because, when harnessed for draft labor, the angle between their line of pull and the direction of traction is smaller. The smaller draft angle also minimizes the uplift on the attached plow and puts less strain on the walking plowman. Lighter animals are also often more agile than heavier ones and may compensate for their lighter weights with tenacity and endurance. And the use of less powerful female animals made sense in all regions where the peasants wanted to minimize the total size of the breeding herd.

But even small and poorly harnessed animals made a great difference for the average peasant (Esmay and Hall 1968; Rogin 1931; Slicher van Bath 1963). A peasant working with a hoe needed at least 100 hours, and up to twice that amount in heavy soils, to prepare 1 hectare of land for planting cereals. Even with a simple wooden plow pulled by a single medium-sized ox, that task could be completed in just over thirty hours. Hoe-dependent farming could have never attained the scale of cultivation made possible by animal-drawn plowing.

Besides speeding up heavy plowing and harvesting, animal labor also made it possible to lift large volumes of irrigation water from deeper wells. And where adequate water and wind resources were either absent or too expensive to harness, animals were used to operate such essential food-processing machines as mills, grinders, and presses at rates far surpassing human capabilities. No less important than the higher output rates was fact that animals gave farmers relief from long hours of tiresome manual labor. These advantages came at a cost—to maintain the animals, the farmers had to produce animal feed. Part of the cultivated land in traditional agricultures had to be set aside to grow hay and coarse grains. This was done easily in North America and in parts of Europe, where the upkeep of horses claimed at times up to one-third of all agricultural area.

Not surprisingly, in China and in other densely populated Asian nations, cattle were the preferred draft animals. Because they are ruminants, they could be maintained solely on roughage from straws and from grazing. They did not have to be fed much grain: Concentrate feed could come largely from such crop-processing residues as brans and oil cakes. Consequently, in China's traditional farming, the cultivation of feed for draft animals claimed only about 5 percent of the annually harvested area.

In India this share was even smaller. Although fodder crops accounted traditionally for about 5 percent of India's cultivated land, most of this feed went to milking animals; some of it ended up nourishing sacred cows (Harris 1966; Heston 1971). Feed for the working bullocks probably claimed less than 3 percent of all farmland. In the most densely populated parts of the Indian subcontinent, cattle survived on a combination of crop by-products, ranging from rice straw and mustard oil cakes to chopped banana leaves, and on whatever they could garner from limited roadside and canal-banks grazing (Odend'hal 1972).

Indian and Chinese draft animals were clear energetic bargains. Many of them did not compete with humans for crop harvests at all; others preempted at most an area of cultivated land that could grow enough food for one person for a year. But the useful annual labor of one of these animals was the equivalent of three to five peasants working 300 days a year! An average nineteenth-century European or American horse could not give such a high relative return but nevertheless provided an energetic boon (see A3.4). Its annual useful labor was equivalent to the work of about six farmers. On average, the land used for feeding each horse (including the nonworking animals) could have grown food for about six people. Even if the nineteenth-century draft horse had served merely as a substitute for tedious human labor it would have earned its keep. But this comparison ignores the fact that strong, well-fed horses could perform tasks beyond human capacity and endurance.

Horses could drag logs and pull out stumps to convert forests to cropland, break up rich prairie soils by deep plowing, or pull heavy machinery. Horses and cattle also produced manure for soil fertilization, milk, meat, and leather. All of these benefits were important. Manure was a source of scarce nutrients and organic matter. In largely vegetarian societies, meat and milk were valuable sources of complete protein. Leather went into making a large number of tools essential to farming and traditional manufactures. And, of course, the animals were self-reproducing.

Irrigation

Diligence has oftentimes, even when nature has failed, availed to bring about the watering of as much land even at the time of smaller rises of the river as at the greater rises, that is, through the means of canals and embankments.

—Strabo, *Geography* (7 b.c.)

The amount of water demanded by various crops depends on many environmental, agronomic, and genetic variables, but the total seasonal need is commonly about one thousand times the mass of the harvested grain. Up to 1500 tonnes of water are needed to grow 1 tonne of wheat; at least 900 tonnes must be supplied for every tonne of rice. About 600 tonnes will suffice for a tonne of corn, the staple grain with the highest water use efficiency (Doorenbos et al. 1979). Thus, for wheat yields of between 1 and 2 tonnes per hectare, a farmer needed a total of between 15 and 30 centimeters of water during a four-month growing season. Annual precipitation in the arid and semi-arid regions of the Middle East, however, ranged from a mere trace to no more than 25 centimeters.

Cropping in such locations thus required irrigation, especially in areas beyond the reach of seasonal floods or in areas with high population densities. Irrigation was also necessary in order to cope with seasonal water shortages. These were especially pronounced in the most northerly reaches of monsoonal Asia, in Punjab, and on the North China Plain. And, of course, rice growing required its own arrangements for flooding and draining the fields.

Gravity-fed irrigation—which uses canals, ponds, tanks, or dams—requires no water lifting and has the lowest energy cost. But in river valleys with minimal stream gradients and on large cultivated plains, it has always been necessary to lift large volumes of surface or underground water. Often the water had to be lifted only across a low embankment into the ridged fields—but no less often it had to be lifted up from steep stream banks or from deep wells.

The amount of energy required for lifting water can be enormous. With weak prime movers and unavoidable inefficiencies, the task could take a very long time. In the absence of lubricants for the simple water-lifting machines, the moving parts did not work smoothly. Irrigation powered by human muscles represented a large labor burden even in societies where tedious work was a norm. Not surprisingly, much ingenuity went into designing mechanical devices using animal power or water flow to ease that work—as well as to make higher lifts possible.

An impressive variety of mechanical devices was invented to lift water for irrigation (Ewbank 1870; Molenaar 1956; Oleson 1984; Fraenkel 1986). The simplest ones—tightly woven or lined shovel-like scoops, baskets, or buckets—were used to raise water less than 1 meter. A scoop or a bucket suspended by a rope from a tripod was slightly more effective. Both of these devices were used in East Asia and the Middle East, but the oldest water-lifting mechanism in widespread use was the counterpoise lift (swape or well-sweep), commonly known as the Arabic *shaduf.*

Recognizable first on a Babylonian cylinder seal of 2000 B.C. and widely used in ancient Egypt (see Figure 3.10), the shaduf reached China by about 500 B.C. and eventually spread throughout the Old World. The shaduf was easily made and repaired: It was basically a long wooden pole that pivoted as a lever from a cross bar or a pole. Its bucket dipper was suspended from the longer arm and counterpoised by a large stone or a ball of dry mud. Its effective lift was usually between 1 and 3 meters. Serial deployment of the device in two to four successive levels was common in the Middle East. Downward pulling on a rope to lift the counterweight could be very tiresome, but cranking an Archimedean screw (Roman *cochlea,* Arabic *tanbur*) in order to rotate a wooden helix inside a cylinder was even more demanding.

Hand- or foot-operated paddle wheels were commonly used in India, Korea, Vietnam, and Japan. In Chinese water ladders (dragon backbone machines: *long gu che*), a series of small boards passing over sprocket wheels formed an endless chain drawing water through a trough. The driving sprocket was inserted on a horizontal pole trodden by two or more men who supported themselves by leaning on a pole.

All of the following devices were powered by animals or by running water. The rope and bucket lift, common in India (*monte* or *charsa*), was powered by one or

FIGURE 3.10 Two water-lifting devices. The counterpoise lift (*top*), common in Egyptian irrigation since 1300 B.C., was worked by just one or two people and could water only small, garden-sized plots. The ancient Chinese dragon backbone machine (*bottom*) was powered by people leaning on a pole and treading a spoked axle. *Sources:* Adapted from Montet (1966) and *Tian gong kai wu* (*The Exploitation of the Works of Nature*; 1637).

two pairs of oxen walking down an incline while lifting a leather bag fastened to a long rope. The Greeks used an endless chain of clay pots on two loops of rope; the pots carried water to the top of a flume. This device, best known as a *saqiya*, spread throughout the Mediterranean with Islam. Its practical lift was limited by the power of the prime mover—usually a single blindfolded animal walking in a circular path. An improved Egyptian version, the *zawafa*, was used for similar lifts, but it delivered more water at a higher rate.

The *noria*, another device widely used in Islamic countries as well as in China, used clay pots, bamboo tubes, or metal buckets fastened to the rim of a single wheel. The wheel could either be driven by right-angle gears and powered by animals or equipped with paddles and powered by water current. This device was rather inefficient because the buckets had to be lifted one wheel radius above the level of the receiving trough. The Egyptian *tabliya* eliminated this problem. This improved device, powered by oxen, included a double-sided all-metal wheel that scooped up water at the outer edge and discharged it at the center into a side trough.

The limits of human performance become obvious when one compares the typical power requirements, lifts, and hourly outputs of these important traditional water-raising devices (see Figure 3.11; A3.5). A single worker could power all Archimedean screws and counterpoise lifts as well as many water ladders. Two men were usually needed to energize high-capacity ladders. Above that, animals were needed. A single small ox could take care of a tabliya or a low-lift saqiya. High-volume lifts of more than 3 meters usually required a pair of animals.

Crop responses to watering can vary so widely that generalizations about the energy returns of traditional irrigation are impossible. Some crops—peanuts, for example—are fairly insensitive; others, like corn, are rather vulnerable. For each crop, the yield response depends on the timing of water availability. It is safe to say, however, that the benefit/cost ratio of crop irrigation was lowest in some parts of the Inca empire. There, irrigation was by gravity flow and no lifting was required, but the construction of canals called for enormous expenditures of human labor.

Some of the canals were surprisingly wide (main lines went up to 10–20 meters); they were carved out of rocks with simple stone tools to carry water over astonishing distances. The main arterial canal between Parcoy and Picuy ran for 700 km to irrigate pastures and fields (Murra 1980). Conquering Spaniards were astonished to see long branch canals carrying water to a small group of cornfields! Naturally, such projects required careful planning and execution in order to maintain appropriate gradients and mobilized large numbers of workers. The pay-off—food energy output from irrigated crops surpassing the huge investment of labor—was obviously postponed for many years, possibly even for decades. Only a confident and well-established central government able to shift resources among different parts of its realm could undertake such an astounding program of public construction.

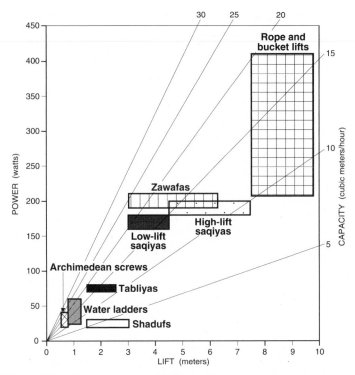

FIGURE 3.11 Lifts, volumes, and power requirements of common water-raising devices and machines. *Source:* Plotted from data in Molenaar (1956), Forbes (1965), and Needham et al. (1965).

In most cases, water management for higher crop yields involved field irrigation, but some agricultures intensified their cropping through an opposite process. In many regions continuous farming would have been unthinkable without the drainage of excessive water. Emperor Yu (2205–2198 B.C.), one of the seven great pre-Confucian sages, owes his place in Chinese history largely to his master plan and heroic organization of a prolonged quest to drain away flood waters (Wu 1982). And the New World's Mayas and successive inhabitants of the Basin of Mexico prospered by adopting intensive forms of crop cultivation that included a variety of water management techniques. These ranged from simple terracing and spring irrigation to elaborate drainage systems and extensive construction of raised fields (Sanders et al. 1979; Flannery 1982).

A unique kind of drainage agriculture evolved over a period of many centuries in part of China's Guangdong province (Ruddle and Zhong 1988). Intensively cropped dikes separated ponds stocked with several species of carp. Recycling of organic wastes—human, pig, and silkworm excreta, crop residues, grasses, weeds, and pond deposits—sustained high yields of mulberries for silkworms, sugar cane, rice, and numerous vegetables and fruits as well as a high output of fish.

Fertilization

Fertilizers for crops: spread pigeon dung on meadow, garden and field crops. Save carefully goat, sheep, cattle, and all other dung … Crops which fertilize land: lupines, beans and vetch. You may make compost of straw, lupines, chaff, bean stalks, husks, and ilex and oak leaves.

—Marcus Cato, *De agri cultura* (second century b.c.)

Atmospheric carbon dioxide and water supply the carbon and hydrogen that form the bulk of plant tissues as carbohydrates. But other elements are absolutely necessary for photosynthesis. Depending on the amount needed they are classified as macro- and micronutrients. The latter are more numerous, including, above all, iron, copper, sulfur, silica, and calcium. There are only three macronutrients: nitrogen, phosphorus, and potassium. Nitrogen is by far the most important one, both because of its biochemical indispensability (it is present in all enzymes and proteins) and because it is the element most likely to be in short supply in continuously cultivated soils (Smil 1985).

In most cases, rain, dust, weathering, and recycling of crop residues replenished withdrawals of phosphorus, potassium, and micronutrients, but continued cropping often created nitrogen deficits. Grains are highly dependent on this nutrient, and shortages cause stunted growth, poor yields, and poor nutritional quality. Traditional farming replaced nitrogen by plowing in straw or plant stalks, recycling animal and human excreta, and cultivating leguminous crops.

Cereal straws were a major potential source of nitrogen, but their direct recycling was limited. Unlike modern short-stalked plants, the traditional cultivated varieties yielded much more straw than grain. Plowing in so much plant mass could have put a great strain on many animals—but such situations almost never arose. Farmers returned only a small fraction of crop residues directly to the soil because they were needed as animal feed and bedding (and thus only recycled in manures), as household fuel, and as a raw material for construction and manufacturing. In wooded regions, straws and stalks were often simply burnt in the field, and thus nitrogen loss was virtually complete.

By recycling urine and excrements as fertilizer, farmers can achieve high crop yields. This practice, perfected over centuries in Europe and East Asia, is limited by the availability of such wastes and requires much repetitive, heavy labor. Handling and treatment of the wastes—cleaning the stalls and sties, allowing the liquids to ferment, composting the mixed wastes—are very time-consuming and unpleasant tasks. Moreover, as most manures have only about half a percent of nitrogen, massive applications were required to meet crops' nitrogen needs. The organic fertilizer that has the highest nitrogen content is guano, the droppings of seabirds preserved in the dry climate of islands along the Peruvian coast. The conquering Spaniards were impressed with its use by the Incas (Murra 1980).

Actual field applications varied widely with the recoverability of the manure (very high with penned animals but negligible with grazers), attitudes toward handling human wastes (ranging from proscriptions to routine recycling), and the intensity of cropping. Much of the nitrogen in recycled waste is lost during the collection, composting, and application process (mainly through ammonia volatilization and by leaching into the groundwater); as a result, the eventual nitrogen uptake by crops is very low (Ross 1989). Because of these losses (which commonly depleted more than two-thirds of the initial nitrogen), traditional farmers had to apply enormous quantities of organic wastes. Consequently, large shares of farm labor had to be devoted to the unappealing and heavy tasks of collecting, fermenting, transporting, and applying organic wastes.

Green manuring was effectively used in Europe ever since ancient times and was also widely employed in East Asia. The practice relied mainly on the special capability of leguminous crops—above all, vetches and clovers. These plants can fix substantial quantities of the nutrient by converting inert atmospheric nitrogen to nitrogen compounds metabolizable by plants. In a mild climate, three to four months of legume growth during the winter added enough nitrogen to the soil to produce a good summer cereal crop.

In areas with high population densities, farmers often must plant yet another food grain crop during the winter months. In the short run this additional crop is advantageous because it produces additional carbohydrates and oils. In the long run, however, it inevitably decreases the total nitrogen availability and affects the yield. Intensive agricultures cannot do without the nitrogen-fixing legumes and must plant them in edible varieties. This desirable practice, repeated every year or as a part of longer crop-rotation sequences, represents perhaps the most admirable energetic optimization in traditional farming. Not surprisingly, it formed the core of all intensive agricultural systems relying on complex crop rotations.

Crop Diversity

Plant millet after millet
And you will end by weeping.

—Peasant saying from
Shandong province of China

Modern commercial agriculture is dominated by monocultures, annual plantings of the same crop. But repeated plantings of the same species have high energy and environmental costs. Monocultures require farmers to use fertilizers to replace the removed nutrients and chemicals to control pests, which thrive on vast uniform plantings. Monocultures of row crops such as corn, which expose much of the soil to rain before the closing of canopies, promote heavy erosion when planted on sloping lands. And constant cultivation of rice in flooded soils deprived of oxygen degrades the soil quality.

In contrast, rotation of cereals and leguminous crops either replenishes the nitrogen or at least eases the drain on nitrogen reserves. Cultivation of a variety of grain, tuber, oil, and fiber crops lowers the risk of total harvest failure, discourages the establishment of persistent pests, reduces erosion, and maintains better soil properties (Lowrance et al. 1984). Rotations can be chosen to fit climatic and soil conditions and dietary preferences. In poor societies rotation can help to ensure food self-sufficiency at the local level.

Although crop rotations are highly desirable from an agronomic point of view, they, too, have their energy cost. Where more than one crop is grown per year (multicropping), the extra tasks obviously require more labor. In places with dry spells, there will be a need for irrigation. And where multicropping is very intensive, with three or even four different species grown every year in the same field, good yields require substantial fertilization. Where two or more crops are grown in the same field at the same time (intercropping), labor demands may be even higher. The fundamental reward of multicropping is that it can support larger populations from the same amount of cultivated land.

The traditional variety of crops and rotation schemes was enormous. For example, in his survey of Chinese farming, John Lossing Buck (1937) counted an astonishing 547 different cropping systems in 168 localities. But several key commonalities are clear. None is more remarkable than the already noted, and virtually global, practice of linking leguminous grains with cereals. Some legumes, most notably soybeans and peanuts, not only contributed to soil fertility and protein supply but also yielded good edible oils. Fats were always welcome in traditional diets, and oilseed cakes and legume stalks were high-protein feeds or fertilizers.

The second commonality was also already noted: that is, rotating green manures with food crops had an important place in every intensive traditional agriculture. A third one was based on producing fibers along with staple carbohydrates (grains, tubers) and oil crops. Consequently, traditional Chinese cropping included numerous schemes of rotating wheat, rice, and barley with soybeans, peanuts, or sesame and cotton and jute. European peasants cultivated fiber plants like flax and hemp in rotation with staple cereals (wheat, rye, barley, oats) and legumes (peas, lentils, beans). Mayan crops included the three staples of New World farming—corn, beans, and squash—but also tubers (sweet potato, manioc, jicama) and fiber plants like agave and cotton (Marcus 1982).

Persistence and Innovation

There is still land suitable for growing grain that has not been brought under
cultivation, resources of hills and lakes that have not been exploited.

—Chao Cuo, *Memorial on the Encouragement*
of Agriculture (178 b.c.)

Traditional farming practices often seem to endure for millennia. This apparent inertia hides numerous but almost always very gradual changes. Innovations can range from better agronomic techniques to new crops. Perhaps the best way to appreciate these changes is to take a long look at the evolution of the four most persistent traditional farming arrangements and then at the rapid advances of North American preindustrial agriculture.

Historically the first agricultural systems developed in the Middle East, were exemplified by Egyptian practices. There the natural limitations (the restricted availability of arable land and the virtual absence of precipitation) and an extraordinary environmental bounty (the Nile's annual floods, which ensured a predictable water and nutrient supply) combined to produce a highly productive agriculture during early dynastic times. At the beginning of the twentieth century, after a long period of stagnation, Egypt still produced some of the highest outputs achievable in solar farming.

Chinese farming is illustrative of East Asia's admirably productive cropping. These practices supported the world's largest culturally cohesive populations, and they survived largely intact until the 1950s. This persistence enabled Western scholars to study them by modern scientific methods and to produce some reliable quantifications of their performance. Complex Mesoamerican societies depended on a unique and highly productive cropping system carried out without plowing and without draft animals. European agriculture evolved from its simple Mediterranean beginnings to a modern system because of rapid advances during the eighteenth and nineteenth centuries. The transfer of European techniques to North America, and the unprecedented rate of innovation during the nineteenth century, created the world's most efficient traditional farming system.

Egypt

I planted for them groves; I dug for them lakes; I founded for them divine
offerings of barley and wheat, wine, incense, fruit, cattle and fowl.

—Ramses III (twelfth century b.c.)

Predynastic Egyptian agriculture, firmly traceable from shortly before 5000 B.C., coexisted with much hunting (of antelopes and pigs as well as crocodiles and elephants), fowling (geese and ducks), fishing (especially easy in flooded shallows), and gathering (herbs and roots). Emmer wheat and two-row barley were the first cereals, and sheep the first domesticated animals. Farmers planted seeds in October and November after the waters of the Nile receded. Crop weeding was rare, and the harvests came after five or six months. The archaeological record indicates that predynastic Egyptian farming could feed perhaps as many as 2.6 people per hectare of cultivated land. A more likely long-term mean was only about half of that rate.

Egyptian farming has always prospered because of irrigation, but in both the Old Kingdom (2705–2205 B.C.) and the New Kingdom (1550–1070 B.C.), irrigation involved relatively simple manipulation of annual floodwaters. Higher and stronger levees were built, drainage channels were blocked off, and flood basins were subdivided (Butzer 1984). Unlike in Mesopotamia or in the Indus valley, in Egypt perennial canal irrigation was not an option: The Nile's very small gradient (1:12,000) made radial canalization impossible. Canal irrigation was first introduced in the Faiyum depression during Ptolemaic times (after 330 B.C.).

Similarly, the absence of effective water-lifting devices greatly limited the dynastic irrigation of higher-lying farmland. Counterpoise lifts, used since the Amarna period in the fourteenth century B.C., were only suitable for irrigating small, garden-like plots. Animal-drawn *saqiyas* needed for continuous high-capacity lifts came only during the Ptolemaic times. Consequently, there was no dynastic cultivation of summer crops, just more extensive winter cropping. Workers harvested the grains, mainly wheat and barley, with wooden sickles set with short notched or serrated flint blades; they cut the straw high above the ground, sometimes just below the heads. This practice, also common in medieval Europe, made for easy harvesting, easier transport of the crop to the threshing floor, and cleaner threshing. In Egypt's dry climate, the standing straw could be cut later as needed for weaving or brickmaking or for use as cooking fuel.

Paintings from Paheri's tomb (fifteenth century B.C.) bring these scenes alive: They depict paired reapers followed by female gleaners, men carrying cut grain in panniers for threshing, and others tossing the grain and chaff into the air to winnow the seed (see Figure 3.12). The accompanying brief inscriptions express eloquently the energetic constraints and realities of the time (James 1984). An overseer prods the laborers: "Buck up, move your feet, the water is coming and reaches the bundles." Their reply, "The sun is hot! May the sun be given the price of barley in fish!" This inscription sums up perfectly both the workers' weariness and their awareness that grain destroyed by water may be compensated for by fish brought by the flood. And the boy whipping the oxen tries to cheer the working animals: "Thresh for yourselves, thresh for yourselves. ... Chaff to eat for yourselves, and barley for your masters. Don't let your hearts grow weary! It is cool" (p. 125).

Besides chaff, oxen were fed barley and wheat straw and grazed on the wild grasses of the floodplain and on the cultivated vetches. As cultivation intensified,

59

FIGURE 3.12 Scenes of farming life from Paheri's tomb. *Source:* James (1984). Reprinted by permission of University of Chicago Press. © 1984 by T.G.H. James.

cattle were seasonally driven to graze in the delta marshes. Farmers harnessed oxen with double-head yokes to do the plowing, and the scattered seed was trampled into the ground by sheep. Records from the Old Kingdom indicate not only large numbers of oxen but also substantial cow, donkey, sheep, and goat herds.

Butzer's (1976) reconstruction of Egypt's demographic history has the Nile valley's population density rising from 1.3 persons per hectare of arable land in 2500 B.C. to 1.8 persons per hectare in 1250 B.C. and to 2.4 persons per hectare by the time of Rome's destruction of Carthage (146–149 B.C.). During the period of Roman rule, Egypt's total cultivated land was about 2.7 million hectares, with about three-fifths of it in the Nile delta. On this land the country's farmers could produce about 1.5 times as much food as was needed for its nearly 5 million people. This surplus was a matter of great importance for the prosperity of the expanding Roman empire: Egypt was Rome's largest grain supplier. Afterwards, during the late Roman, Byzantine, and early Islamic eras, Egyptian farming declined and stagnated.

Even as recently as the second decade of the nineteenth century Egypt was cultivating only half as much land as it did during Roman rule. But because of higher yields, this agriculture supported about as many people as the area twice as large had nearly two millennia before. Productivity rose rapidly only with the spread of perennial irrigation after 1843, when the first Nile barrages provided adequate waterheads to feed canal networks. The national multicropping index rose from just 1.1 during the 1830s to 1.4 by 1900 (Waterbury 1979). During the 1920s the multicropping index surpassed 1.5. Farming was still powered by animate energies but, already helped by inorganic fertilizers, Egyptian peasants were feeding 6 people from every hectare of cultivated land.

China

> *Heavily flapping are the bustards in row:*
> *They have perched on mulberry.*
> *The king's business never ends:*
> *I cannot plant my millet;*
> *How can my parents have a taste?*
>
> —*Shi jing* (*Book of Poetry*) (eleventh century b.c.)

China underwent long periods of turmoil and stagnation, but its traditional farming was considerably more innovative than Egypt's (Bray 1984; Ho 1975). As elsewhere, the early stages of Chinese farming were not at all intensive. Before the third century B.C. there was no large-scale irrigation and little or no double-cropping and crop rotation. Dryland millets in the north and rainfed rices in the lower Yangzi basin were the dominant crops. Pigs were the most abundant and oldest domestic animals, but clear evidence of manuring dates only after 400 B.C. By the time Egypt was supplying the Roman empire with grain, the Chinese had developed several tools and practices that Europe and the Middle East adopted only centuries, or even more than a millennium, later.

These advances included, above all, iron moldboard plows, collar harnesses for horses, seed drills, and rotary winnowing fans. All these innovations came into widespread use during the early Han dynasty (207 B.C.–A.D. 9). Perhaps the most important was the widely adopted cast-iron moldboard plow. Made from non-brittle metal (casting was perfected by the third century B.C.), these mass-produced plows extended the possibilities of cultivation and eased the heavy work. Although heavier than wooden plows, they created much less friction and could be pulled by a single animal even in the water-logged clay soils of the South.

Multitube planting drills reduced the seed waste inevitable with hand broadcasting, and crank-operated winnowing fans greatly shortened the time needed to clean threshed grain. The efficient collar harness for horses did not actually make that much of a difference. In the North, the less demanding oxen were not displaced by horses, and only water buffaloes, harnessed by neck yokes, could be used in the wet fields of the South.

No other dynastic period can compare with the Han in terms of fundamental changes in farming. Subsequent advances were slow, and after the fourteenth century rural techniques were nearly stagnant. Grain output increased between the Ming dynasty (1368–1644) and the early Qing dynasty (1644–1911), but more than half of this increase was due to the expansion of the cultivated area. Higher labor inputs—above all more irrigation and fertilization—accounted for most of the remaining increase. Better seeds and new crops, most notably corn, made some regional difference.

Undoubtedly China's most important and lasting contribution to intensified cropping was its design, construction, and maintenance of extensive irrigation systems. Nearly half of all projects operating by 1900 had been completed before 1500 (Perkins 1969). The most famous one, Sichuan's Dujiangyan, which still waters fields growing food for several tens of millions of people, goes back to the third century B.C. Min Jiang's bed was cut at the river's entrance to the plain at Guanxian. The stream was then repeatedly subdivided by rocky arrowheads built into the midflow. The water was diverted into branch canals, and the flow was regulated by dykes and dams. Baskets of woven bamboo filled with rocks constituted the main building ingredient. Dredging and dyke repairs during low-water seasons have kept this irrigation system working for more than 2000 years.

The construction and unceasing maintenance of such irrigation projects (as well as the building and dredging of large ship canals) required long-range planning, massive mobilization of labor, and major capital investment. None of these requirements could be met without effective central authority. There was clearly a synergistic relationship between China's impressive large-scale water projects and the rise, perfection, and perpetuation of the country's hierarchical bureaucracies.

Human-powered water lifting was tedious and time consuming, and its energy costs were rather high—but so were the rewards of higher yields. When irrigation supplied additional water during critical growth periods, its net energy return was easily around 30 units of food energy gained for every unit of energy expended (see A3.6). During less sensitive growth periods, irrigation could still return around 20 times as much food energy in the additional harvest as the peasants consumed in treading water ladders.

Manure and nightsoil applications during the late nineteenth and the early twentieth centuries commonly averaged 10 tonnes per hectare in the rice region. Rates of up to four times as much were not uncommon. Huge amounts of organic wastes were brought from cities and towns, which created a large waste-handling and transportation industry. The intensity of Chinese manuring was admired by Western travelers who were curiously unaware of how closely it was matched by earlier European experience (King 1927). But no culture surpassed the applications of organic wastes in South China's Guangdong dike-and-pond region: between 50 and 270 tonnes of pig and human excrements per hectare (Ruddle and Zhong 1988).

Composting and regular applications of other organic wastes—ranging from silkworm pupae to canal and pond mud and from waterweeds to oilseed cakes—further increased the burden of collecting, fermenting, and distributing. Not surprisingly, at least 10 percent of all labor in traditional Chinese farming was devoted to management of fertilizers. In the North China Plain, heavy fertilization of wheat and barley was commonly the single most time-consuming task for both humans and animals (for the former, it represented nearly one-fifth of the labor devoted to those crops; for the latter, about one-third). But this investment was very rewarding: Its net energy return was commonly over fifty-fold (see A3.7).

The overall food energy returns in traditional Chinese cropping were not so high even during the early decades of the twentieth century. The main reasons were that there was little mechanization of cropping processes and a continuing reliance on human labor. Because of a wealth of quantitative information on virtually all aspects of traditional Chinese farming in the 1920s and 1930s (Buck 1930, 1937), it is possible to describe the system in great detail and construct accurate energy accounts.

The typical size of fields was very small (only about one-fifth of a hectare) and the fields were, on average, just a five- or ten-minute walk from a farmhouse. Nearly half of the farmland was irrigated, and a quarter of it was terraced. More than 90 percent of the cropped land was planted in grains, less than 5 percent with sweet potatoes, 2 percent with fibers, and 1 percent with vegetables. Only about one-third of all northern farms had at least one ox, and less than a third of southern holdings owned a water buffalo. Except for plowing and harrowing, Chinese field work relied almost exclusively on human labor. As both oxen and water buffaloes were fed hardly any grain, the energy returns can be calculated just on the basis of the detailed human labor budgets available for all major food crops.

Unirrigated northern wheat yielded usually no more than 1 tonne per hectare. Its production required more than 600 hours of labor, and it returned between 25 and 30 units of food energy in unmilled grain for every unit of food energy needed for field work and crop processing. Local and regional rice yields were already rather high during the Ming dynasty, and the national average was around 2.5 tonnes per hectare during the early decades of the twentieth century, second only to Japan. About 2000 hours of labor were needed to produce such harvests; thus, gross energy returns were between 20 and 25 units. Gross energy returns for corn were up to 40 units, but cornmeal was never a favored food. For legumes (soybeans, peas, beans) the returns were rarely higher than 15 units, and commonly just around 10. That was also the return for plant oils pressed from rapeseed, peanuts, or sesame seeds.

Ancient Chinese population densities could not have been much different from the Egyptian means; they probably ranged from just around 1 person per hectare in the poorest northern regions to well over 2 people per hectare in the southern rice areas. Gradual intensification of cropping, combined with simple diets, did eventually support much higher rates. Perkins (1969) estimated about 2.8 people per hectare of farmland in 1400 and posited that this figure had risen to 4.8 people per hectare by 1600. During the prosperous Qianlong reign (1736–1796), the density declined slightly because an expanded population was able to open up new farmland.

Densities again increased during the nineteenth century. By 1900 the average rate was above 5 people per hectare—higher than the contemporary mean for Java and at least 40 percent above the Indian mean. Buck's (1937) surveys indicate a national average of at least 5.5 people per hectare of cultivated land in China in the 1930s. This density nearly matched that of Egypt for the same decade, but in Egypt all the land was irrigated and inorganic fertilizers were already being used. In contrast, China's output was depressed by northern dryland farming. Large parts of the South were supporting more than 7 people per hectare by the late 1920s. In comparison to dryland wheat cropping, wet rice farming invariably had lower net energy returns. These returns were more than counterbalanced by greater yields per hectare. Double-cropping of rice and wheat in the most fertile areas could yield enough to feed 12–15 people per hectare.

Mesoamerica

At the time when they are to prepare their fields for sowing, they first make a prayer to the land, telling it that it is their mother and that they want to open her. … At this time they seek the favor of Quetzalcoatl in order that he give them strength to be able to till the land.

—Don Pedro Ponce, *beneficiado* of Tzumpahuacan (1629)

Without any plowing or draft animals, the agricultures of classical Mesoamerican civilizations differed substantially from those of Old World cultures. But they, too, evolved more intensive cropping methods to support impressively high population densities. The tropical Mayan lowlands and the much drier high-lying Basin of Mexico were the regions that made the greatest achievements. Although there was a great deal of interaction between these cultures, and although corn was the staple crop for both, their histories were substantially different. The first culture succumbed to internal problems that have not yet been fully explained; the other was destroyed by Spanish invasion.

Mayan culture developed gradually for a long time before the beginning of its classic period in about A.D. 300 (Hammond 1986). The Mayan region, encompassing parts of today's Mexico (Yucatan), Guatemala, and Belize, sustained a complex civilization until about A.D. 1000. Then, in one of the most enigmatic episodes of world history, the classic Mayan society disintegrated and its population nearly disappeared, declining from about 3 million people during the eighth century to just around 100,000 by the time of the Spanish conquest (Turner 1990). Agricultural malpractice—above all, excessive erosion and breakdown in water management—has been suggested as one of the reasons for the Mayan collapse, but there is no clear evidence to support such a conjecture.

In the early stages of their development Mayans were shifting cultivators, but they turned gradually to intensive forms of cropping (Flannery 1982; Turner 1990). Upland Mayas built extensive rock-wall terraces, which conserved water and prevented heavy erosion on continuously cultivated slopes. Lowland Mayas built some impressive canal networks and raised fields above the floodplain level in order to prevent seasonal inundation. Ancient elevated ridged fields, some dating from as early as 1400 B.C., are still discernible on modern aerial photographs. By identifying these fields and accurately dating them, in the 1970s scholars were able to disprove the long-prevailing notion that the Mayas only practiced shifting agriculture (Harrison and Turner 1978).

The Basin of Mexico saw a succession of complex cultures, including the Teotihuacanos (100 B.C.–A.D. 850), the Toltecs (960–1168), and, after the early fourteenth century, the Aztecs (Tenochtitlan was founded in 1325). These changes were accompanied by a long transition from plant gathering and deer hunting to settled farming. Intensification of cropping through water regulation started early in the Teotihuacan era; by the time of the Spanish conquest it had evolved to such a degree that at least one-third of the region's population depended on water management for its food (Sanders et al. 1979).

Permanent canal irrigation around Teotihuacan enabled the land to support about 100,000 people, but the most intensive cultivation in Mesoamerica was on the *chinampas*. These rectangular fields were raised to between 1.5 and 1.8 meters above the shallow waters of the Texcoco, Xalco, and Xochimilco Lakes. Excavated mud, crop residues, grasses, and water weeds were used in their construction. The farmers cropped their rich alluvial soils continuously, or allowed only a few months of rest, and reinforced their edges with trees.

Chinampas turned unproductive swamps into high-yielding fields and gardens and solved the problem of soil waterlogging. Easy accessibility by canoes made for relatively effortless transportation of grain and produce to large urban markets. Given the high yields of chinampa cultivation, these raised fields provided an outstanding return on the invested labor. The food energy consumed in building a chinampa could be returned in increased yields in less than a year (see A3.8). This high benefit/cost ratio explains the frequency of the practice, which started as early as 100 B.C. and reached its peak during the last decades of Aztec rule.

At the time of the Spanish conquest the Texcoco, Xalco, and Xochimilco Lakes had about 12,000 hectares of chinampas (Sanders et al. 1979). Their construction required at least 70 million days' worth of labor. The average peasant could not work more than about 100 days annually on large hydraulic projects. A good portion of this time had to be devoted to the maintenance of existing embankments and canals; thus, seasonal labor of at least 60 and up to 120 peasants was needed to add 1 hectare of new chinampas. The means were different—but the pre-Hispanic Basin of Mexico was clearly as much a hydraulic civilization as Ming China, its great Asian contemporary. Long-term, well-planned, centrally coordinated effort and an enormous expenditure of human labor were the key ingredients of its agricultural success.

Irrigated corn is an inherently higher-yielding crop than wheat, and population densities supported by the best Mesoamerican farming were very high. A hectare of a high-yielding chinampa could feed as many as 13–16 people deriving 80 percent of their food energy from the grain. Naturally, averages for the whole Basin of Mexico were considerably lower, ranging from less than 3 people per hectare in fringe areas to about 8 people per hectare in areas with well-drained soils and permanent irrigation (Sanders et al. 1979). The basin's preconquest (1519) population of around 1 million people, using all cultivable land in the valley, constituted an average density of about 4 people per hectare. Nearly identical densities were supported by raised-field cultivation of potatoes in the wetlands around Lake Titicaca, the core area of the Incas between today's Peru and Bolivia (Denevan 1982).

Europe

> *Between this place and Chatillon sur Indre I saw several women spreading dung with their hands! ... The principal difference between the ploughs, carts, wagons, harrows, and the like in France, and such implements in England, is that these things seem to be ... about the same as ... used in England a great many years, perhaps a century, ago.*
>
> —James Paul Cobbett, *A Ride of Eight Hundred Miles in France* (1824)

In Europe, much as in China, periods of relatively steady improvement alternated with stagnation in productivity and major regional peacetime famines persisted until the nineteenth century. But until the seventeenth century, European farming was generally inferior to Chinese accomplishments, and Europe was always belatedly adopting innovations coming from the East. Greek farming, of which we know little, was certainly not as impressive as Middle Eastern farming. Romans had gradually evolved a moderately complex agriculture that exerted a significant influence up to the seventeenth century (White 1970; Fussell 1972). The works of Cato, Columella, Palladius, Pliny, Varro, and Virgil contain interesting references to Roman farming techniques.

Unlike farming in densely populated China, European agriculture always had a strong component of animal husbandry. Roman mixed farming included rotations of cereal and legume crops, composting, and plowing-in of legumes as green manure. All possible organic wastes, ranging from the highly valued pigeon excreta to oil cakes, were recycled. Farmers reduced soil acidity through repeated liming of their fields (chalk or marl applications). At least one-third of the fields lay fallow at any given time.

Oxen, often shod, were the principal Roman draft animals. Plows were wooden, sowing was by hand, and harvesting was done with sickles. A mechanical Gallic reaper, described by Pliny and pictured on a few surviving reliefs, was in limited use. Threshing was done by animals or with flails, and yields were low and highly variable. In Roman wheat farming during the first few centuries A.D., between 180 and 250 hours of human labor (and around 200 hours of animal labor) were required to produce a typical harvest of about half a tonne of grain per hectare. Gross energy returns were between 30 and 40 units (see A3.9).

The productivity of European farming changed little during the millennium between the demise of the Western Roman Empire and the beginnings of the great European expansion. In the early thirteenth century, wheat production proceeded by means largely unchanged since Roman times and could not support population densities higher than the predynastic Egyptian average. But the Middle Ages were definitely not a period devoid of notable technical innovation (Duby 1968; Fussell 1972; Lizerand 1942; Seebohm 1927; Slicher van Bath 1963). Perhaps the most important change was the adoption of the shoulder collar for draft horses.

Largely because of this improved harness, horses started to replace oxen as the principal draft animals in the richer regions of the continent. But the transition took centuries to accomplish. In the better-off regions of Europe, this change occurred between the eleventh century and the sixteenth century. It is well documented that in England horses made up only 5 percent of all demesne draft animals but about 35 percent of the draft animals on peasant holdings at the time of the Domesday count in 1086 (Langdon 1986). By 1300 these shares had risen, respectively, to 20 and 45 percent, and after a period of stagnation, horses constituted the majority of draft animals by the end of sixteenth century.

The relative richness of English data also illustrates the complexity of this transition. For a long time horses were simply substituted for oxen as pacesetters in mixed teams. Their adoption had a clear regional pattern (East Anglia was far ahead of the rest of England), and smallholders were much more progressive in using them on their farms. Several factors determined the pace of the transition in different regions. These included differences in prevailing soil types (clays favored oxen), the availability of feed (extensive pastures favored oxen), and accessibility to markets where good working animals could be bought and meat sold (proximity to towns favored horses). No less important was the conservatism of the elites and their resistance to change (illustrated by the slow rate of change on demesnes), on the one hand, and a pioneering spirit and the interest in striving for lower operating costs and gaining prestige, on the other.

The transition was further hindered by poorly designed plows and by the relative weakness of most medieval horses. The combination of wide wooden soles, heavy wooden wheels, and large wooden moldboards resulted in enormous friction. In wet soils it was not uncommon to use four to six oxen or horses to overcome this resistance. Moreover, farmers had to make frequent stops to clear compressed weeds and soil clogging the gap and the angle between the share and the flat moldboard.

In spite of their inefficiency, the flat moldboard plows were essential to extending the land under cultivation, as were larger animal teams (increasingly including horses). By dividing fields into raised lands and sunken furrows, moldboard plowing created conditions for effective artificial drainage. Although certainly much less spectacular than chinampas, this form of controlling excessive field water had widespread spatial and historical repercussions. Moldboard plowing opened up the extensive waterlogged plains of northern Europe to the cultivation of wheat and barley, crops native to dry Middle Eastern environments.

By the late Middle Ages the frontier of German settlement marked the easternmost extent of moldboard plowing. The technique encompassed all European flatlands between the North Sea and the Urals only by the nineteenth century. Clearly, its use was both a revolutionary change assuring agronomic advances in this region as well as a key ingredient of the continuing agricultural prosperity of cool, wet lowlands.

Heavy draft horses, common on European farms and roads during the nineteenth century, were a product of many centuries of breeding (Villiers 1976). Draft horse weight and power began to rise appreciably only after the thirteenth and fourteenth centuries, when Europeans started to breed warhorses heavy enough to carry armored knights. These changes spilled over to benefit the working stock, and heavier draft horses became common after 1400. Still, these animals did not match the size and weight of later horses. Animals taller than 16 hands and weighing close to 1 tonne did not emerge for another three or four centuries (see Figure 3.13). This fact explains the medieval English complaints that horses were useless in heavy clay soils. In contrast, the heavy draft horses of the nineteenth century were outstanding on wet land, in heavy soils, and on uneven ground.

FIGURE 3.13 European draft horses ranged from small pony-like animals (less than 12 hands, or 1.2 meters, tall) to huge, massive beasts (taller than 16 hands and weighing around 1 tonne). *Source:* Based on Silver (1976).

During the nineteenth century a pair of good horses could easily do 25 to 30 percent more field work in a day than a team of four oxen. This speed enabled farmers to cultivate existing fields more frequently (and to plow fallow land to kill weeds), to extend cropping to new land, and to spend more time on other field or farmyard activities. And in most European regions, crop rotations could provide enough concentrate feed to make the maintenance of a two-horse team cheaper than the upkeep of four oxen.

Because the rate of transition from oxen to horses was slow, farm output underwent many major regional fluctuations and very low staple grain harvests persisted for too long; in other words, productivity did not steadily improve as the number of draft horses grew. The superiority of draft horses became obvious only when the more powerful animals made up the majority of the draft stock and started to work as part of a much more intensified farming during the seventeenth and eighteenth centuries. In road transport, the advantages of horses were recognized much earlier.

Working horses also posed a major energy supply challenge to traditional farming. The heavy work possible with efficient harnessing and horseshoes had to be energized by better feed than the roughages (grasses or straws) sufficient for working cattle. Powerful working horses needed concentrates like cereal or legume grains. Consequently, farmers had to intensify their cropping to provide for both their families and their animals. Multicropping and crop rotations diffused gradually at first, and intensive agriculture was born in regions with population densities that were still low; without the need for animal feed, these methods would not have become established until much later.

The abundance of historical price figures makes it possible to reconstruct long-term production trends for a number of countries (Abel 1980). Naturally, there were substantial regional differences, but large-scale cyclical fluctuations are unmistakable. Times of relative prosperity (most notably 1150–1300, the sixteenth century, and 1750–1800) were marked by extensive conversions of wetlands and forests to fields. Prosperity also spurred the colonization of remote areas and brought a greater variety of foods to supplement the staple of all European diets—bread.

Periods of major decline—the fourteenth century, the depression around 1500, and much of the seventeenth century—brought famines, large population losses, and extensive abandonment of fields and villages (Beresford and Hurst 1971; Centre des Recherches Historiques 1965). Epidemics and wars were behind most of the great population losses of the fourteenth century. In the early decades of the fifteenth century Europe had about one-third fewer people than in 1300, and Germany lost about two-fifths of its peasants between 1618 and 1648. Some European populations continued to decline even after the Thirty Year War.

European peasant farming could not assure a secure food supply right up to the end of the eighteenth century. Only then did a fairly intensive cultivation finally become the norm. Its foundations were laid by a diffusion of new agricultural societies and reforms. Farmers gradually abandoned fallowing and adopted several standard crop rotations. The cultivation of potatoes became widespread after 1770, livestock production expanded, and heavier manuring became regular. In eighteenth-century Flanders, annual applications of manure, night soil, oil cakes, and ash easily averaged 10 tonnes per hectare (Slicher van Bath 1963).

The Netherlands emerged as the leader in crop productivity. In the 1800s, wheat became the principal food crop on Dutch farms. Barley, oats, rye, beans, peas, potatoes, rapeseed, clover, and green fodder were also important crops (Baars 1973). The best wheat yields were about four times higher than in the Middle Ages, although the total amount of human labor remained roughly the same (see A3.9). Less than 10 percent of the farmland lay fallow, and the field cropping was closely integrated with livestock production.

Farming intensification continued in most European countries after the recovery from an overproduction-induced depression in the early nineteenth century. Two German examples illustrate these changes (Abel 1980). In 1800 about a quarter of German fields were fallowed, but the share was less than 10 percent by 1883. The average annual per capita meat consumption was less than 20 kilograms before 1820, but it was almost 50 kilograms by the end of the century. Earlier three-crop rotations were replaced by a variety of four-crop sequences. In a popular Norfolk cycle, wheat was followed by turnips, barley, and clover, and six-crop rotations were also spreading. Applications of calcium sulphate, and of marl or lime to correct excessive soil acidity, became common in the better-off areas.

Many new farm implements with better designs appeared during the nineteenth century; these inventions diffused rapidly and led to the current era of widespread mechanization. By the middle of the century, yields were rising in every important farming region as the rapidly intensifying agriculture was able to supply food for growing urban populations. After centuries of fluctuations, population densities were rising steadily. In the most intensively cultivated regions of the continent—in the Netherlands and in parts of Germany, France, and England—they reached 7 to 10 people per hectare of arable land by the year 1900. But these levels already reflected considerable energy support received indirectly through machinery and fertilizers produced with coal and electricity. European farming in the late nineteenth century became a hybrid energy system: While still critically dependent on animate prime movers, it was increasingly benefiting from the many inputs of fossil energy.

North America

> *The Moldboard of this well, and so favorably known PLOUGH, is made of wrought iron, and the share of steel, 5/16th of an inch thick, which carries a fine sharp edge. The whole face of the moldboard and share is ground smooth, so that it scours perfectly bright in any soil, and will not choke in the foulest of ground.*
> —John Deere's advertisement of his new plough (1843)

The history of postrevolutionary American farming is remarkable because of the rapid, ever accelerating rate of innovation. These changes resulted in the world's most labor-efficient crop cultivation by the end of the nineteenth century (Ardrey 1894; Rogin 1931; Schlebecker 1975). During the last decades of the eighteenth century, America lagged behind Europe in terms of cropping innovations and other advances, especially in the South. Wooden plows had wrought-iron shares and wooden moldboards that were shoddily covered with metal pieces; they caused high friction and heavy clogging and put a strain on the yoked oxen. There were no efficient harrows; farmers sowed by hand, harvested by sickle, and, in the Northeast, threshed by flailing (in the South the threshing was done even more primitively by animal treading).

All of this changed rapidly in the nineteenth century. Changes in plowing came first. Jethro Wood's patents (of 1814 and 1819) made interchangeable cast-iron moldboards practicable. Subsequent improvements in durability and design were brought into general use by the early 1830s. Improved cast-iron plows were soon superseded by steel plows. The first of these was made from saw-blade steel by John Lane in 1833. John Deere began commercial production of steel plows in the 1840s; Lane made further improvements and introduced a layered steel plow in

1868. Two- and three-wheel riding plows followed after 1864 (see Figure 3.14). Gang plows, with up to ten shares and drawn by as many as a dozen horses, were being used before the end of the century. Massive steel moldboard plows made it possible to cut the heavily sodded grasslands and open up North America's vast plains for grain cropping.

Advances in plowing were matched by other innovations. Seed drills and horse-powered threshing machines were widely used by 1850. The first mechanical grain reapers were patented in England between 1799 and 1822; two American inventors, Cyrus McCormick and Obed Hussey, built on this basis to develop practical mass-produced machines starting in the 1830s (McCormick 1931). They started to sell heavily during the 1850s, and 250,000 of them were in use by the end of the Civil War. The first harvester, patented in 1858 (by C. W. and W. W. Marsh), required only two men to bind the cut grain. After several decades of unsuccessful attempts by many inventors, John Appleby introduced the twine knotter in 1878.

This invention was the last ingredient needed for a fully mechanical grain harvester that could discharge tied grain sheaves ready for stacking. The rapid diffusion of these machines before the end of the nineteenth century, together with gang plowing, made it possible for settlers to open up huge expanses of grasslands, not only in North America but also in Argentina and Australia. But the performance of the best twine-binding harvester was soon surpassed by the first horse-drawn combines, which were introduced by California's Stockton Works during the 1880s. Housers, the company's standard combines after 1886, cut two-thirds of California's wheat by 1900. The largest ones needed up to 40 horses to operate and could harvest 1 hectare of wheat in less than 40 minutes.

The leaps in productivity from 1800 to 1900 were impressive. In 1800, seeding by hand and working with ox-drawn wooden plows and brush harrows, sickles, flails, and winnowing sheets, New England farmers spent 150 to 170 hours of labor to produce a good harvest of wheat. By 1900, using horse-drawn gang-plows, spring-tooth harrows, and combines, California farmers could produce the same amount of wheat in less than nine hours of labor (see A3.10). In 1800, New England farmers needed more than 7 minutes of labor to produce 1 kilogram of wheat; in 1900, California's Central Valley farmers required less than half a minute to get the same return. This example shows roughly a twenty-fold improvement of labor productivity.

In terms of net energy expenditures the differences were even more impressive: Most of the hours in 1800 were spent in much heavier labor (with walking plows, scythes, and flails) than in later decades. Moreover, seeding and storage losses declined appreciably with the new techniques. In 1900, every unit of food energy needed for farm labor produced, on the average, about twenty-five times more edible energy in wheat grain than in 1800. Naturally, these huge advances were only partially due to better machinery. They also resulted from the substitution of horse power for human muscles. American inventors produced a vast range of ef-

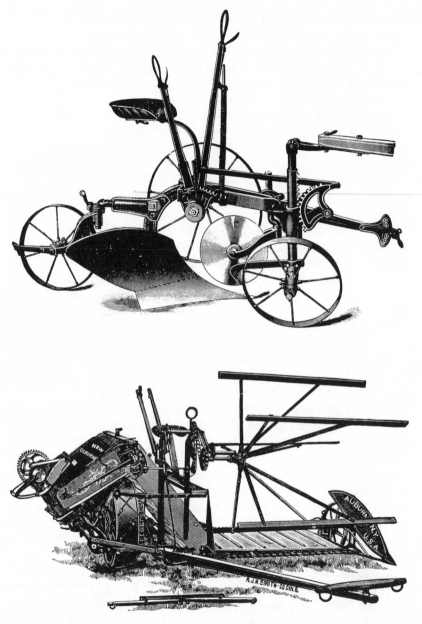

FIGURE 3.14 The three-wheeled steel riding plow (*top*) and the twine harvester (*bottom*). These two innovations opened up the American plains for large-scale grain cropping. The plow illustrated here was made by Deere & Co in Moline, Illinois, during the 1880s; the harvester is an Osborne model made during the last decades of the nineteenth century in Auburn, New York. *Source:* Reproduced from Ardrey (1894).

ficient implements and machines, but they had only a limited success in displacing draft animals as prime movers.

Threshing was the only major operation where horses were gradually replaced by steam engines. America's rapidly expanding agriculture relied on growing stocks of horses and mules. These were generally powerful, large, well-fed animals—and their energy costs were surprisingly high. Horses helping to produce California wheat in 1900 consumed about 50 percent more energy than New England oxen in 1800—and while they were working that energy had to come in the form of oats or corn, not just hay and straw. Growing this feed grain reduced the production of crops for humans.

It is possible to calculate the feeding requirements of American horses and mules fairly accurately (USDA 1959). During the first two decades of the twentieth century the numbers of American horses and mules averaged about 25 million heads. Feeding them required about one-quarter of America's cultivated land (see A3.11). This huge claim was possible only because of America's plentiful farmland. In 1910 the country had almost 1.5 hectares per capita, twice as much as in 1990, and about ten times as much as in China during the early 1900s.

During the last decades of the nineteenth century, it was not just clever designs and plentiful horse power that made American farming so productive. During the 1880s American coal consumption surpassed wood combustion and crude oil was starting to gain in importance. Coal and oil fueled the production and distribution of tools, implements, and machines and made it possible to ship agricultural products. American farmers ceased to be just skilled managers of renewable solar flows: Their outputs were subsidized by fossil fuels.

Limits of Traditional Farming

Even a clever daughter-in-law
Cannot cook a meal without rice.
—**Chinese proverb**

The enormous socioeconomic changes that can take place over centuries sometimes make it easy to forget that both the prime movers and basic cropping practices remained largely unchanged across millennia of preindustrial history. Traditional methods thus had definite limits, even with more efficient use of animate power and intensified cropping practices. In such highly productive regions as northwestern Europe, central Japan, and the coastal provinces of China, yields approached these limits by the end of the nineteenth century. Preindustrial agricultures brought only very limited improvements in average harvests and provided no more than basic subsistence diets for most of the population, even in good years. Extensive malnutrition, and in some regions even recurrent famines, were still unavoidable.

Achievements

The country people are producing vast quantities of supplies for food, shelter,
clothing, and for use in the arts. The country homes are improving. ... As
agriculture is the immediate basis of country life, so it follows that the general
affairs of the open country, speaking broadly, are in a condition of improvement.

—U.S. Country Life Commission, *Report* (1909)

Advances in traditional farming came slowly, and adoption of new methods did not mean a general disappearance of old practices. Fallow fields, scythes, and inefficiently harnessed oxen did not disappear in late nineteenth-century Europe when annual cropping, grain harvesters, and good horse teams became common. The only way to reduce human labor in a nonmechanized system was through a more widespread use of draft animals. This shift required not only better harnessing, feeding, and breeding methods but also innovative designs for field tools and machines.

The rate of advance accelerated during the eighteenth century. The progress in American wheat cultivation cited above provides a good example. Although the substitution of animal draft for human labor increased productivity, for a long time these gains were barely discernible. Accurate information is scarce—we have no reliable national or regional data for periods before the early nineteenth century—but it is clear that stagnation and marginal gains were the norm in both Europe and Asia. The available data show that, during the Middle Ages, because of weather extremes, yields could even be lower than the seeded amount!

Some estimates suggest that average returns in the early Middle Ages were just twofold for wheat. The best available long-term reconstructions of a national trend are those documenting English yields over the past seven centuries (Bennett 1935; Stanhill 1976; Clark 1991). In the thirteenth century, English wheat seed returns ranged mostly between 3 and 4 units, with recorded maxima up to 5.8. These figures translate to a mean of just above half a tonne per hectare. Careful analyses of all available English evidence show that this very low yield was irreversibly doubled only some five centuries later. English wheat yields remained at nearly medieval levels until about 1600, but afterwards they steadily increased. The countrywide mean for 1500 was doubled by the middle of the nineteenth century, largely as a result of extensive land drainage and the widespread adoption of crop rotation and intensive manuring (see Figure 3.15).

At that time British agriculture was already benefiting from much-improved machinery and rapid advances in the nation's economy fueled by an increase in coal combustion. The effect of these fossil energy subsidies is also clearly discernible in the case of Dutch harvests, which, in the latter half of the eighteenth century, amounted to as much as 15–20 times the seed sown. Dutch yields averaged 1 tonne per hectare nationwide by the year 1800. In contrast, French wheat yields showed only a mild upward trend even during the nineteenth century, and in

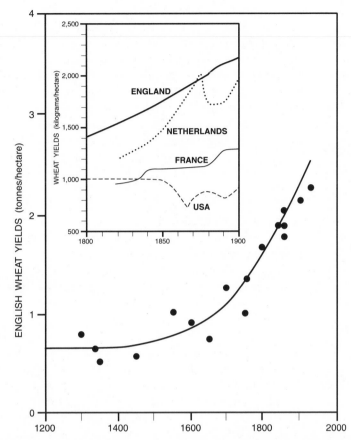

FIGURE 3.15 English wheat yields show a long period of stagnation followed by a post-1600 takeoff. Yield gains during the nineteenth century were even more impressive in the Netherlands, but marginal in France. In the United States the westward expansion of wheat farming to the drier interior actually resulted in declining average yields. *Sources:* Based on data in USDA (1955), U.S. Bureau of the Census (1975), Mitchell (1975), Stanhill (1976), and Clark (1991).

America, despite gains in labor efficiency, yields actually declined with the expansion of cropping on Western plains.

The best available data on yield averages show that during medieval times, an hour of labor produced no more than 3 to 4 kilograms of grain. By 1800 the average rate was around 10 kilograms per hour. A century later it was close to 40 kg and the best performances yielded more than 100 kg. Energy returns increased a bit faster: An average hour of field work in the late nineteenth century called for less physical exertion than a typical hour of labor in medieval times—hand plowing with a heavy wooden moldboard plow pulled by oxen required much greater effort than riding on a steel plow pulled by a team of powerful horses.

A complete wheat-growing sequence in late Roman or early medieval times yielded roughly a forty-fold net energy gain in the harvested grain. At the beginning of the nineteenth century, good Western European harvests returned about 200 times more energy in wheat than they consumed in production. By the century's end, the ratio was commonly above 500, and the best returns were above 2500. Net energy gains (after subtracting seed requirements and storage losses) were necessarily lower, no more than 25 times the energy input for usual medieval harvests, between 80 and 120 times the energy input by the beginning of the nineteenth century, and typically between 400 and 500 times the energy input by the end of the century.

These soaring labor productivities were bought at a price, namely, the growing deployment of draft power and a substantial energy investment in animal feeding. In the Roman case, every unit of useful power available from human work was supplemented by about 8 units of animal labor capacity. In early nineteenth-century Europe, the typical ratio of human to animal power capacity rose only to around 1:15—but on the most productive American farms it was well above 1:100 during the 1890s. Human labor became a negligible source of mechanical energy. Farmers' work shifted almost completely to management and control, tasks that required little power but resulted in high outputs.

The energy costs of draft power increased rapidly. A pair of Roman oxen subsisting on roughage did not need any grain to perform field tasks, but a medium-sized pair of early nineteenth-century European horses consumed almost 2 tonnes of feed grain per year, about nine times the amount of food grain consumed by people per capita. During the 1890s, a dozen powerful American horses typically consumed some 18 tonnes of oats and corn per year, about eighty times the amount eaten by their master. Only a few land-rich countries could provide so much feed. Feeding twelve horses would have required about 15 hectares of farmland. An average U.S. farm had almost 60 hectares of land in 1900, but only one-third of it was cropland. Clearly, even in the United States, only large grain growers could afford to keep a dozen or more working animals: The 1900 average was only three horses per farm (U.S. Bureau of the Census 1975).

Not every traditional society could intensify its farming by relying on higher inputs of animal labor. Cropping intensification based on more elaborate cultivation of a limited amount of arable land became the norm in Asia. The most notable examples of this development were in Japan, parts of China and Vietnam, and Java, the most densely settled island of the Indonesian archipelago. This approach, which Clifford Geertz (1963) aptly called "agricultural involution," rested on the high yield potential of irrigated rice and on the heavy investment of energy that went into constructing and maintaining irrigation systems, wet fields, and terraces over centuries.

Although cropping intensification in dryland farming can easily lead to environmental degradation (especially soil erosion and nutrient loss), paddy agroecosystems are much more resilient. Their assiduous cultivation is an enormous absorber of human labor. Farmers must carefully level their fields, sprout seedlings in nurseries, and micromanage the planting, weeding, and harvesting processes. Once established, this introversive tendency is difficult to break. The process supports progressively higher population densities but leads eventually to extreme impoverishment. Labor productivity first stagnates then starts declining as larger populations rely on increasingly marginal diets. Many regions of China showed clear signs of agricultural involution during the Ming and Qing dynasties.

After the conflicts of the first half of the twentieth century, Maoist policies based on mass rural labor in communal farming perpetuated the involution until the late 1970s. At that time 800 million peasants still represented more than 80 percent of China's total population, and they continued to subsist on barely adequate, although more equitably distributed, rations. Only Deng Xiaoping's abolishment of communes and de facto privatization of farming during the early 1980s radically reversed the trend. In contrast, Japan broke the trend with the Meiji restoration in 1868. Between the early 1870s and 1940, Japan's total population, and its rice yields, more than doubled, while the rural population declined by half to just 40 percent of the total (Taeuber 1958).

In spite of their fundamental differences, the two grand patterns of farming intensification—the substitution of animal for human labor and the maximization of peasant labor inputs—both slowly pushed up agricultural production. This process was essential to the higher densities and growing complexity of urban civilizations. It released a growing share of labor for nonagricultural work and led to occupational specialization.

These changes can be reconstructed only in approximate terms. Past population totals are highly uncertain even for societies with a long tradition of relatively comprehensive counts (Durand 1960; Whitmore et al. 1990). Reliable data on cultivated land, and on shares of land planted to annual or permanent crops, are even more scarce. Consequently, it is impossible to present reliable trends of population densities. It is possible, however, to contrast the minima characteristic of early agricultures with a number of typical later performances (derived from written records) and then with the best achievements of the most intensive preindustrial farming methods, which are well documented by modern research.

The average rates for all ancient civilizations appear to start around 1 person per hectare of arable land (see Figure 3.16). Only after many centuries of slow advances did this rate double. In Egypt the doubling took about 2000 years, and it appears that a very similar amount of time was needed in both China and Europe. Subsequent intensification of farming produced much faster growth rates. The best regional achievements peaked at more than 10 people per hectare of arable land. But abstract rates should not be the primary concern in such comparisons: Nutritional adequacy and variety should also be considered.

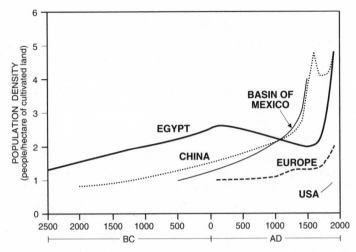

FIGURE 3.16 Approximate long-term trends of population densities per hectare of farmland in Egypt, China, the Basin of Mexico, and Europe. *Sources:* Based on estimates and statistics in Perkins (1969), Mitchell (1975), Butzer (1976), Waterbury (1979), Richards (1990), and Whitmore et al. (1990).

Nutrition

He ... asserts they use little or no milk or potatoes, and no oatmeal. Seldom any butter, but occasionally a little cheese and sometimes meat on Sunday. ... Bread, however, is the chief support of the family, but at present they do not have enough, and his children are almost naked and half starved.

—**Sir Frederic Morton Eden,** *The State of the Poor* (1797)

It is impossible to calculate the average food requirements of a traditional society with any reasonable certainty. Production estimates must rely on cumulative assumptions, and actual consumption was affected by high, and highly variable, post-harvest losses. Perhaps the only acceptable generalization, based on documentary and anthropometric evidence, is that there is no clear upward trend in per capita food supply across the millennia of traditional farming. Some early agricultural societies were in certain aspects relatively better-off, or at least no worse-off, than their successors. For example, Ralph Ellison's (1981) reconstruction of ancient Mesopotamian ration lists indicates that daily energy supplies between 3000 and 2400 B.C. were about 20 percent above the early twentieth-century mean for the same region.

Calculations based on records of the Han dynasty show that in the state of Wei during the fourth century B.C., a typical peasant was expected to provide each of his five family members with nearly half a kilogram of grain a day (Yates 1990). This total is identical to the North Chinese mean during the 1950s before the introduction of pumped irrigation and synthetic fertilizers (Smil 1981). More reliable figures for early modern Europe also show some notable declines of staple food consumption even in cities enjoying privileged food deliveries. For example, the yearly per capita supply of grain in the city of Rome fell from 290 kilograms during the late sixteenth century to about 200 kg by the year 1700. The average per capita availability of meat also declined, from almost 40 kg to just around 30 kg (Revel 1979).

In most cases, the more recent diets were also less diverse. They contained less animal protein than the earlier diets, mainly because the latter included a larger variety of wild animals, birds, and aquatic species. This qualitative decline was not offset by a more equitable availability of basic foodstuffs: Major consumption inequalities, both regional and socioeconomic, persisted until the nineteenth century. Large portions of traditional farming societies, and at times even majorities, had to live with food supplies below the levels necessary for a healthy and vigorous life. For example, in Eastern Prussia, one-third of the rural population could not afford enough bread as late as 1847 (Abel 1980).

Even during fairly prosperous times when typical diets supplied more than adequate nutrition in terms of total energy and basic nutrients, foods were highly monotonous and not very palatable. In large parts of Europe, the everyday staples were dark bread, coarse grains (oats, barley, buckwheat), turnips, cabbage, and, later, potatoes. These ingredients were often combined in thin soups and stews, and evening meals were indistinguishable from breakfasts and midday foods. Typical rural Asian diets were, if anything, even more likely to be dominated by a few cereals. In premodern China, millet, wheat, rice, and corn supplied more than four-fifths of all food energy. The Indian situation was almost identical.

Seasonally abundant vegetables and fruits enlivened this monotony. Asian favorites included cabbage, radishes, onions, garlic, and ginger as well as pears, peaches, and oranges. Cabbage and onions were also among the mainstays in Europe, along with turnips and carrots, apples, pears, plums, and grapes. The most important Mesoamerican species were the tomato, chayote, chile, papaya, and avocado. These foodstuffs provided only a small amount of food energy but were essential sources of vitamins and minerals.

Typical rural diets in Asia were always overwhelmingly vegetarian, as were those of the Mesoamerican societies, which, with the exception of dogs, never had any large domesticated animals. But some parts of Europe enjoyed relatively high meat intakes during prosperous periods. Still, most Europeans had only an occasional small serving of meat. Animal proteins were consumed largely in the form of dairy products. Roasts, stews, beer, cakes, and wine were common only during such festive occasions as religious holidays, weddings, or guild banquets.

Even when everyday diets supplied sufficient energy and protein, they were frequently deficient in vitamins and minerals. Mesopotamian diets, although rich in barley, were short in vitamins A and C: Ancient inscriptions carry references to blindness and a scurvy-like disease (Ellison 1981). During subsequent millennia these two deficiencies were common in most extratropical societies. Very low intakes of meat and leafy green vegetables caused chronic iron deficiencies. Rice-dominated diets caused major calcium deficits, especially among children: In South China the mean was less than half of the recommended daily intake (Buck 1937). Monotonous and inadequate diets and widespread malnutrition remain a norm today in many poor countries where population densities have surpassed the limits sustainable by even the most intensive traditional farming.

Limits

We must not deceive ourselves. A great change in the agricultural position is impending.

—J. Caird, *The Landed Interest and the Supply of Food* (1880)

In spite of its slow progress, traditional farming was an enormous evolutionary success. There would have been no complex cultures without the high population densities supported by permanent cropping. Even an ordinary staple grain harvest could feed, on the average, ten times as many people as the same area dedicated to shifting cultivation. But there were clear limits to the population densities achievable with traditional farming. Moreover, the average food supply was rarely much above the existential minimum, and seasonal hunger and recurrent famines weakened even the societies with rather low population densities, good soils, and relatively good farming techniques.

Energy supply was the most common limit to the process of substituting animal draft for human labor. The production of the concentrate feed needed for working animals could not be allowed to compromise adequate harvests of food grains. Even in land-rich agricultures with extraordinary feed production capacities, the trend of substitution could not have continued much beyond the American achievements of the late nineteenth century. Heavy gang plows and combines took animal-drawn cultivation to its practical limit. The time was ripe for a much more concentrated, and much more powerful, prime mover—the internal combustion engine.

Societies undergoing agricultural involution were able to subsist on gradually diminishing per capita returns of human labor. In these cultures, population densities were eventually limited by the maximum possibilities of nitrogen recycling. The most intensive application of traditional nitrogen sources—largely from the recycling of organic wastes and the planting of green manures—provided enough of the nutrient to support 12–15 people per hectare of cultivated land. This limit could not be surpassed by traditional farming methods. Manure production could not be increased beyond the limit set by the availability of animal feed. Moreover, heavy manuring and use of human wastes is quite taxing in terms of strenuous, repetitive labor.

The only universally available, effective alternative was to rotate the crops demanding fertilization with leguminous plants. Yet this is also a limited solution. Frequent planting of leguminous green manures would maintain high soil fertility—but it would inevitably lower the average annual output of staple cereals. Leguminous grains can be cultivated largely without an external supply of nitrogen, and they are high in protein, but they also have disadvantages. They are difficult to digest and often have low palatability. Moreover, they cannot be used to bake bread or, with a few exceptions, to make noodles. Unfailingly, as societies become more prosperous their consumption of legumes declines.

Regardless of the historical period, the environmental setting, or the prevailing mode of cropping and intensification, no traditional agriculture could consistently produce enough food to eliminate extensive malnutrition. All these cultures were vulnerable to major famines. Droughts and floods were the most common natural triggers. In China in the 1920s, peasants could recall an average of three crop failures brought about by such disasters that had occurred within their lifetime and that were serious enough to cause famine (Buck 1937). These famines averaged about ten months and forced a quarter of the affected population to eat bark and grasses. Nearly one-seventh of all people left their villages in search of food. A similar pattern would likely be found in most Asian and African societies.

Some famines were so devastating that they remained in collective memory for generations and led to major social, economic, and agronomic changes. Notable examples of such events include the frost- and drought-induced failures of corn harvests in the Basin of Mexico between 1450 and 1454 (Davies 1987); the famous collapse of the *Phytophthora*-infested Irish potato crops between 1845 and 1852 (Edwards and Williams 1957); and the great drought-induced famine of 1876–1879 in India (Seavoy 1986).

To better insulate themselves from the recurrence of such drastic food shortages, these preindustrial societies would have had to extend their cultivated land, intensify their cropping, or both—and they were constantly trying to do so. But in an overwhelming majority of cases these moves were undertaken reluctantly, and they were commonly postponed for so long that natural disasters were repeatedly translated into major famines.

The reason for this tardiness is clear: Extended and intensified cultivation required a higher investment of energy. Even in societies that could afford a large number of draft animals, most of the additional energy inputs had to come from longer hours and harder exertion of human labor. Moreover, intensified food production often had a lower energy benefit/cost ratio than its less intensive predecessors. Not surprisingly, traditional cultivators tried to postpone these greater labor burdens and lower relative returns. Usually they expanded or intensified their cropping only when they were forced to do so in order to provide for the basic needs of their gradually increasing populations. In the long run this reluctant expansion and intensification could support substantially larger populations—but per capita food availabilities, as well as the quality of average diets, hardly changed for centuries or even millennia.

Less energy-intensive agricultural practices were persistent. The transition from shifting cultivation to permanent cropping usually took a very long time, and peasants were reluctant to expand farming into new areas or to intensify their cropping. When gradual population increases could not be sustained by local or regional production, peasants were more likely to extend the cultivated area than they were to intensify the cropping of existing land. Consequently, it took centuries or millennia to adopt annual cropping instead of extensive and prolonged fallowing.

There is no shortage of historical examples illustrating this reluctance. Shifting cultivation in forest environments offered basic subsistence and meager material possessions—but for many societies it remained the preferred way of life even after many generations of contact with permanent farmers. Sharp contrasts between the alluvial farmers and mountain dwellers could be seen well into the twentieth century in southern Chinese provinces, throughout Southeast Asia, and in many parts of Latin America and sub-Saharan Africa. And the practice was persistent even in Europe.

In the Ile de France, the fertile region surrounding Paris, shifting cultivation (with fields abandoned after just two harvests) was still common in the early twelfth century. And on the margins of the continent, in Finland and in northern Russia, shifting cultivation was practiced as late as the first two decades of the twentieth century. Lowland peasants were reluctant to expand cultivation to marginal mountain or wetland soils.

The villages of Carolingian Europe were overpopulated and their grain supplies constantly insufficient—nevertheless, except in parts of Germany and Flanders, there was little effort to create new fields beyond the most easily cultivable lands (Duby 1968). Later European history is replete with waves of German migrations from the densely populated western regions. Armed with superior moldboard plows, these migrants opened up farmlands in areas considered inferior by local peasants in Bohemia, Poland, Romania, and Russia—and set the stage for violent nationalist conflicts for centuries to come.

Opening up new farmland required additional labor, but in most instances this one-time energy investment was a fraction of the investment needed for multicropping, manuring, terracing, irrigation, ditching, or field raising. And so, even in the relatively densely populated regions of Asia and Europe, it took millennia to move gradually from extensive fallowing to annual cropping and multicropping. In China, every dynasty adopted in its early years a policy of extending the cultivated land as the primary means of feeding a growing population (Perdue 1987). In Europe, fallowing was still common as recently as the beginning of the seventeenth century; as much as 35 to 50 percent of the land lay fallow at any given time. And the more intensive triennial system coexisted in England with the biennial one from the twelfth century to the seventeenth; the triennial system finally prevailed only during the eighteenth century (Titow 1969).

Not surprisingly, the transition from shifting cultivation to permanent farming and its subsequent intensification usually happened first in areas with poorer soils, limited arable land, high aridity, or uneven precipitation. Environmental stresses and high population densities certainly did not cause every instance of cropping intensification, but a strong relationship is unmistakable. An excellent early example of such intensification comes from the archaeological findings in northwestern Europe. There is clear evidence that the transition from the Neolithic to the Bronze Age started first in areas with limited arable land, in today's Switzerland and Britain (Howell 1987).

The archaeological evidence also indicates that intensification in Yucatan Maya started first in the environments that were either more marginal (drier) or more fertile (hence more densely settled) than the average locations (Harrison and Turner 1978). The historical record also makes it clear that this pattern is a common one.

For example, Hunan province, with its good alluvial soils and usually abundant precipitation, is now by far the largest rice producer in China. Yet, in the early fifteenth century, it was still a sparsely populated frontier—even though the dry and erosion-prone valley of the Wei He (the site of Xi'an, China's ancient dynastic capital) had turned to intensive farming more than a millennium earlier. And farmers in the densely settled Flanders were one to two centuries ahead of most of their German or French counterparts in reclaiming wetlands and intensifying fertilization (Abel 1980).

These examples document a fundamental energetic preference: Traditional farming (or peasant) societies prefer to minimize the labor needed to secure a basic food supply and essential material possessions. Traditional peasants behaved as gamblers. They tried to stay on slim margins of food surplus for too long, betting that the weather would cooperate and help to produce another fair harvest next year. But given the low staple grain yields and relatively high seed to harvest ratios, they lost repeatedly, and often catastrophically.

Ronald Seavoy (1986) labeled this behavior a "subsistence compromise." But he insisted that the sole reason for this compromise was that peasants universally preferred indolence—that this preference, indeed, was a primary social value in peasant societies. I find this hypothesis unacceptable. Similarly, Gregory Clark (1987), looking at agricultural productivity in the United States and Britain, on one side, and in central and eastern Europe, on the other, attributed the substantial differences almost solely to faster work rates in the two English-speaking nations.

Such sweeping generalizations ignore the influence of many critical factors. Environmental conditions—soil quality, the amount and reliability of precipitation, the per capita availability of land, fertilizer, and food, and the capability to support draft animals, for example—have always influenced productivity. So have socioeconomic peculiarities (land tenure, corvée, taxation, tenancy, ownership of animals, and access to capital) and technical innovation (better agronomic methods, animal breeds, plows, and cultivation and harvesting implements). John Komlos (1988) considered some of these factors in his persuasive refutation of Clark's exaggerations.

Undoubtedly, many cultures did put a low social value on the physical labor of cultivation, and there were important differences in work rates among traditional agricultures. But these realities arose from a complex combination of social and environmental factors, not merely from simple indolence on the part of subsistence peasants or motivation and hard work on the part of other farmers driven by a desire to accumulate wealth.

A much less contentious generalization concerning physical work is that it was spread among as many people as possible. In practice this principle meant that much of the work was transferred to women and children, generally persons of lower status in peasant societies (Caldwell 1976). Women were responsible for a high share of field and household tasks in virtually every traditional society. And because even pregnancy and lactation did not add much of a burden in terms of additional food, and because the children often started working as early as four or five years of age, large families were the least energy-intensive way to minimize adult labor and secure food in old age when infirmities set in.

In traditional agricultures powered overwhelmingly by human labor it was clearly rational to minimize individual workloads by having large families. At the same time, this strategy made it much more difficult to increase average per capita food availability and avoid recurrent famines. Only the inputs of fossil energies—directly as fuels and electricity and indirectly in agricultural chemicals and machinery—could sustain both an expanding population and a higher per capita availability of food.

APPENDIXES

A3.1   Labor and Energy Requirements in Traditional Farming

Tasks	People/Animals	Hours per Hectare	Energy Cost
Hoeing			M–H[a]
General	1/—	100–120	M–H
Wet soil	1/—	150–180	H
Plowing			M–H
Wooden plow	1/1	30–50	H
Wooden plow	1/2	20–30	H
Steel plow	1/2	10–15	M
Harrowing	1/2	3–10	M
Sowing			L–M
Broadcasting	1/—	2–4	M
Seed drills	1/2	3–4	L
Weeding	1/—	150–300	M–H
Harvesting			M–H
Sickle (wheat)	1/—	30–55	H
Sickle (rice)	1/—	90–110	H
Cradle	1/—	8–25	H
Binding sheaves	1/—	8–12	M–H
Shocking	1/—	2–3	H
Reaper	1/2	1–3	M
Binder	1/3	1–2	M
Combine	4/20	2	M
Threshing			L–H
Treading	1/4	10–30	L
Flailing	1/—	30–100	H
Threshers	7/8	6–8	M
Winnowing	1/—	20–30	M–H

[a]Light work consumes less than 20 kJ of food energy per minute for an average adult man. Moderate exertions range up to 30 kJ/minute, and the heavy ones up to 40 kJ/minute. Analogical rates for women are about 30 percent lower.

Sources: Ranges were compiled and calculated from data in Bailey (1908), Rogin (1931), Buck (1937), Shen (1951), and Esmay and Hall (1968). Energy cost indicators were estimated from metabolic studies reviewed in Durnin and Passmore (1967).

A3.2 Typical Energy Expenditures in Traditional Farming

Common tasks in traditional farming required at least moderate expenditures of energy during the course of a working day. The typical energy needs of these moderate activities were about four times the basal metabolic rate, or 1000 and 1300 kilojoules per hour (kJ/ hour) for women and men, respectively. One must subtract the number of kilojoules usually taken up by basic existential needs to find the net labor energy cost, which averages roughly 800 kJ/hour. I will use this figure to represent the net food energy cost of an average hour of labor in traditional agriculture.

A3.3 Typical Weights, Drafts, Working Speeds, and Power of Common Domestic Animals

	Weights (kg)		Typical	Usual	Power
Animals	Common Range	Large Sizes	Draft (kg)	Speed (m/s)	(W)[a]
Horses	350–700	800–1000	50–80	0.9–1.1	500–850
Mules	350–500	500–600	50–60	0.9–1.0	500–600
Oxen	350–700	800–950	40–70	0.6–0.8	250–550
Cows	200–400	500–600	20–40	0.6–0.7	100–300
Buffaloes	300–600	600–700	30–60	0.8–0.9	250–550
Donkeys	200–300	300–350	15–30	0.6–0.7	100–200

[a]Power values are rounded to the nearest 50 W.

Sources: Based on Hopfen (1969), Rouse (1970), and Cockrill (1974).

A3.4 Energy Cost, Efficiency, and Performance of a Draft Horse

Maintaining an average workhorse weighing 500 kilograms required about 70 megajoules (MJ) of digestible energy a day (Brody 1945). Actual daily rations, made up of a varying proportion of highly digestible grains and less digestible crop straw and hay, were about 25 percent higher than this amount. Feed requirements during the labor periods were usually between 1.5 and 1.9 times the maintenance need. With draft equal to 15 percent of its body weight (75 kg), a horse moving at the speed of 1 meter per second (m/s) could do about 20 MJ of useful work in eight hours. Its all-day energy efficiency would thus average around 13 percent (20/150).

The average energy efficiency of moderately hard human labor was also about 13 percent: The similarity is not surprising—the same biochemical imperatives govern the functioning of these two working mammals. Although efficiencies are comparable, overall performances are not. People cannot sustain useful power rates of more than 70 to 100 watts (W), while draft animals will work for hours at rates of 500 to 800 W. Except for donkeys, animals are thus worth at least 5 or 6, and commonly about 8, laboring men. And during recurrent periods of heavy field work, well-fed animals can work for days at rates equivalent to the labor of 13 to 15 men.

A3.5 Power Requirements, Lifts, Capacities, and Efficiencies of Traditional Water-Lifting Devices

Devices	People/ Animals	Lift (m)	Capacity (m³/h)	Work (kJ)	Input Efficiency[a] (kJ)	(%)
Scoops	2/—	0.6	5	30	440	7
Suspended scoops	2/—	1	8	80	440	18
Shaduf	1/—	2.5	3	75	220	34
Archimedean screw	2/—	0.7	15	100	440	23
Paddle wheel	1/—	0.5	12	60	220	27
Water ladder	2/—	0.7	9	60	440	14
Rope and bucket	3/4	9	17	1500	5690	26
Saqiya	1/2	6	8	470	2740	17
Zawafa	1/2	6	12	710	2740	26
Noria high lift	1/2	9	9	790	2740	29
Noria low lift	1/1	1.5	22	325	1480	22
Tabliya	1/1	2.5	12	295	1480	20

[a]Energy costs were calculated by assuming an average power input of 60 W for people and 350 W for draft animals.

Sources: Compiled and calculated from data in Molenaar (1956), Forbes (1965), and Needham et al. (1965).

A3.6 Energy Benefit/Cost Ratio of Irrigation

Field studies have shown that winter wheat yields decline by about half when a 20 percent shortfall in annual water supply occurs during the critical flowering period (Doorenbos et al. 1979). A good late Qing dynasty harvest of 1.5 tonnes per hectare (t/ha) would thus be reduced by about 150 kg on a typical small field of 0.2 ha. Making up the deficit of 10 centimeters of rain through irrigation would require 200 tonnes of water—but the actual supply from a canal had to amount to twice that mass. This was because of irrigation efficiency: The share of delivered water that plants actually used was typically only 50 percent in simple ridge-and-furrow irrigation. The other half of the irrigation water was lost to soil seepage and evaporation.

Using a traditional Chinese water ladder, two peasants could lift 400 tonnes of water about 1 meter in approximately 80 hours. This work would require about 65 megajoules (MJ) of additional food energy, and the increased wheat yield would contain about 2 gigajoules (GJ) of digestible energy (including a 10 percent reduction to account for seed and storage losses). Consequently, water ladder treading returned about 30 times more food energy than it consumed.

A3.7 Energy Benefit/Cost Ratio of Fertilization

A good late Qing dynasty winter wheat harvest of about 1.5 tonnes per hectare (t/ha) required just over 300 hours of human labor plus about 250 hours of animal labor. Fertilization made up 17 percent and 40 percent of these totals, respectively. I assume, conservatively, that the 10 tonnes of fertilizer applied per hectare contained only 0.5 percent nitrogen (Smil 1985). After inevitable losses (mainly leaching and volatilization), only half of this amount would actually be available to the crop. Each kilogram of nitrogen would result in additional production of about 10 kilograms of grain.

Thus, a fertilized crop yielded at least 250 kg more grain than an unfertilized crop. No more than 3 to 4 percent of this grain was used as animal feed. After milling, the food grain yielded at least 200 kg of flour, or about 2.8 gigajoules (GJ) of food energy—compared to an investment of about 40 megajoules (MJ) of additional food for human labor. Every unit of food energy invested in fertilization thus returned around 70 units of edible crop, an impressive benefit/cost ratio.

A3.8 Raised Fields in the Basin of Mexico

An excellent chinampa corn harvest of 3 tonnes per hectare (t/ha) produced about 30 gigajoules (GJ) more food energy than a dryland plot (including a 10 percent reduction to account for seed and storage losses). Fields were raised at least 1.5 meters above the water level, so 1 hectare of chinampas required a build-up of some 15,000 cubic meters (m³) of lake silt and mud. A man working five to six hours a day would emplace no more than 2.5 m³. Constructing 1 hectare of chinampas would thus require about 6000 man-days of labor. Assuming a labor energy cost of 900 kilojoules per hour (kJ/h), the task would call for about 30 GJ of additional food energy—an amount gained in increased harvests in just a single year.

A3.9 Labor Requirements and Energy Costs of European Wheat Harvests, 200–1800

	Hours of Labor (people/animals) Spent in Farming One Hectare of Wheat		
Tasks	Roman Italy 200	England 1200	Netherlands 1800
Plowing			
Oxen	37/74	25/150	
Horses			15/30
Harrowing	8/16	7/14	5/10
Sowing			
Broadcasting	4/—	4/—	
Seed drill			3/6
Manuring			40/60
Harvesting			
Sickle	50/—	50/—	
Cradle			24/—
Hauling	15/30	10/20	7/14
Threshing			
Treading (oxen)	30/60		
Flailing		30/—	33/—
Winnowing	25/—	25/—	30/—
Measuring, sacking	8/—	7/—	10/—
Hours of labor	177	158	167
Energy cost (MJ)	142	126	134
Grain yield (t/ha)	0.4	0.5	2.0
Food yield (GJ)	3.3	4.9	22.2
Net energy return	23	39	166
Hours of animal work	180	184	120

Sources: Calculations are based on information in Baars (1973), Seebohm (1927), White (1970), Stanhill (1976), and Langdon (1986).

A3.10 Labor Requirements and Energy Costs of U.S. Wheat Harvests, 1800–1900[a]

Tasks	1800	1850	1875	1900
Plowing				
Wooden plow	20/40			
Cast iron plow		15/30		
Steel plow			8/24	
Steel gang plow				3/30
Harrowing				
Brush harrow	7/14			
Tooth harrow		5/10	5/15	0/4
Seeding				
Broadcasting	3/—			
Drilling		3/6	3/9	1/2
Harvesting				
Sickle	49/—			
Cradle		25/—		
Binder			11/6	
Combine				3/17
Hauling	10/10	8/8	5/5	2/10
Threshing				
Flailing	33/—			
Threshers		10/10	8/8	
Winnowing	40/—			
Hours of labor	162	66	40	9
Energy cost (MJ)	145	56	32	7
Gross food energy return	129	335	586	2680
Net food energy return	90	270	500	2400
Labor productivity (minutes/kg of grain)	7.2	2.9	1.8	0.4
Hours of animal labor	64	64	67	63

[a]Accounts and balances were prepared for four representative cases. The first one (1800) is a typical New England cultivation where two oxen and one to four men powered all the tasks. The second sequence (1850) traces inputs in horse-powered mid-century farming in Ohio. The third one (1875) shows further advances in Illinois, and the last account (1900) reviews the most productive form of U.S. horse-powered wheat growing, in California. Figures in the table are total hours (men/animals) spent per hectare of wheat growing. Because the yields of U.S. wheat did not show any upward trend during the nineteenth century, I assume a constant yield of 20 bushels per acre, or 1350 kilograms of grain (18.75 GJ) per hectare.

Source: Accounts are based largely on performance rates compiled by Rogin (1931).

A3.11 Feeding America's Draft Animals

American farmers owned a total of 25 million horses and mules at the beginning of the twentieth century. Two-thirds of these, or 16.5 million animals, were actually working. If the average daily intake was 4 kg of grain for working animals and 2 kg of concentrate feed for the rest (Bailey 1908), the annual need was roughly 30 million tonnes of oats and corn. Because the prevailing grain yields were about 1.5 tonnes per hectare (t/ha), farmers would have had to plant at least 20 million hectares of feed grains.

To get adequate roughage, working horses needed at least 4 kilograms of hay a day; non-working animals could be maintained with about 2.5 kilograms a day. Thus, all animals combined required an annual total of roughly 30 million tonnes of hay. Average hay yields were about 3 t/ha, so at least 10 million hectares of hay had to be harvested. Thus, the total area devoted to horse feed had to be no less than about 30 million hectares, compared to around 125 million hectares of annually harvested land. Consequently, America's farm horses required almost 25 percent of the country's cultivated land.

4

Preindustrial Prime Movers and Fuels

During unrecorded millennia the muscles of men and animals had supplied the necessary motion for the earliest anonymous machines—the quern, the potter's wheel, the bow drill, the forge bellows, or the hand pump. In time men learned to use the power of fire and water to ease the onerous tasks.

—Louis C. Hunter, *A History of Industrial Power in the United States, 1780–1930*

MOST PEOPLE in preindustrial societies spent their lives as peasants and used farming methods and tools much like those of their ancestors. Nevertheless, their labor and occasional advances sufficed to support gradually larger and more complex urban societies. Their achievements enabled them to construct remarkable structures, to increase the capacity and reach of transportation, and to make numerous improvements in metallurgy and manufacturing techniques. Although the prime movers and fuels energizing these advances remained unchanged for millennia, human ingenuity improved the performance of several prime movers in many remarkable ways. Eventually, some of these conversions became so powerful and so efficient that they helped to energize the initial stages of modern industrialization.

Increases in output and efficiency have two main causes. The first is the multiplication of small forces—primarily a matter of superior organization of massed animate labor. The second is technical innovation. In practice, the two approaches were often combined. For example, the monumental structures built by virtually every high old culture demanded both massed labor and extensive applications of labor-easing devices ranging from levers, inclined planes, and pulleys to cranes, windlasses, wheelbarrows, and treadwheels.

The differences between the first clearly documented mechanical energy convertors and their successors at the beginning of the industrial era are often stunning. For example, the spoon-tilt hammer, the simplest machine powered by falling water, did not even involve continuous rotary motion: It was merely a simple lever, an exact opposite of the counterpoise lift. In contrast, the nineteenth-century water-powered forge hammer was an impressive, complex, high-performance piece of machinery.

Similar comparisons can be made for every class of water- and wind-powered prime mover. What a difference there is, for example, between the rough-hewn horizontal wooden waterwheels of medieval times, capable of delivering less than half a horsepower, the seventeenth-century vertical machines, capable of producing ten times that amount, and the Lady Isabella, England's largest iron overshot wheel, which had a capacity equivalent to that of more than 500 strong horses! Or between inefficient medieval post mills, which had to be laboriously turned into the wind only to lose more than four-fifths of their potential power to poor sails and rough gearing, and automatically regulated nineteenth-century mills, which had spring sails and smooth transmissions.

The contrasts are no less impressive for animate conversions and the combustion of biomass fuels. A heavy nineteenth-century draft horse hitched to a light, flat-topped wagon on a hard-topped road, equipped with iron horseshoes and a collar harness, could easily pull a load twenty times heavier than the load that could be pulled by its much lighter, unshod, breast-harnessed ancestor linked to a heavy wooden cart on a muddy road. And an eighteenth-century blast iron furnace consumed less than one-tenth of the charcoal per unit of hot metal output than was needed by its early medieval predecessor.

Human exertions, however, changed little between antiquity and the centuries immediately preceding industrialization. Average body weights hardly increased. All the essential devices providing humans with a mechanical advantage have been with us since the time of the ancient empires or even before that (or how else did Stonehenge's 40-tonne outer stones get raised?).

In this chapter I will first appraise the traditional prime movers as well as the combustion of biomass fuels. Afterwards I will look in some detail at applications in critical segments of traditional economies: food preparation, provision of heat and light, land- and water-borne transportation, construction, and metallurgy.

Prime Movers and Biomass Fuels

Slowly the water-wheel and the windmill are improved by the craftsmen, but those prime-movers do not become significantly stronger until the production of better gear-wheels and the careful measurement of the efficiency of the various types of prime movers.

—R. J. Forbes, *Studies in Ancient Technology* (1965)

Animate labor and the kinetic energies of water and wind were the only prime movers before the diffusion of industrial civilization fueled by coal. Biomass fuels—wood, charcoal, crop residues, and dung—provided all household and' industrial heat. For most of the Western world the dominance of muscles, water, wind, and wood came to an end only in the latter half of the nineteenth century. Even for the industrialized countries, the traditional prime movers and fuels are thus more than a matter of passing historical interest: They formed the foundation of our present affluence.

In many parts of the world, the poor still rely heavily on animal and human muscles and on wood, crop residues, or dried dung. For them, these sources of energy remain existential necessities. This continuing dependence marks the great divide between the rich and poor. But it also allows scholars to study these energy conversions with modern scientific methods in an effort to understand the long history of preindustrial practices.

Animate Power

Oxen—weary; man—hungry; the sun already high.

—Bo Zhuyi,
The Charcoal-seller (early ninth century)

This scene from Tang poetry was repeated countless times in human history: miserable roads, tired animals, hungry men, unfinished tasks. Animate energies remained the most important prime mover for most of humankind until the middle of the twentieth century. The limited power of humans and animals, circumscribed by metabolic requirements and mechanical properties, restricted the reach of preindustrial civilizations. Societies deriving their kinetic energy solely or overwhelmingly from animate power could provide most of their members with little physical security or material affluence.

There were only two practical ways in which to increase the delivery of useful animate power: either by concentrating individual inputs or by using mechanical devices to amplify muscular exertions. The first approach soon runs into practical limitations, especially with direct deployment of human muscles. Even an unlimited labor force is of little use for directly grasping and moving a relatively small but very heavy object: Only a limited number of people can fit around its perimeter. And although a group of people may be able to carry a heavy object, lifting it first in order to insert slings or poles could be an insurmountable problem.

Individuals can only lift and move loads that are substantially lighter than their own bodies. Without the help of mechanical devices, many essential construction and transportation tasks would have been impossible. The three simplest aids—levers, inclined planes, and pulleys—were used by virtually all old high cultures (Lacey 1935; Needham et al. 1965; Burstall 1968). Common variations and combinations of these basic tools included wedges, screws, wheels, treadwheels, and gearwheels. By using these tools and simple machines, people were able to deploy

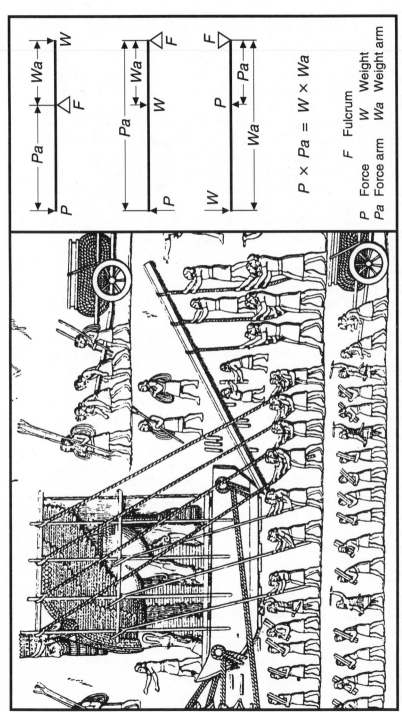

$$P \times Pa = W \times Wa$$

F Fulcrum

P Force	W Weight
Pa Force arm	Wa Weight arm

FIGURE 4.1 Three classes of levers, distinguished by the point at which the force is applied in relation to the object (whose weight, W, always acts in downward direction) and the fulcrum. In the first type of lever (*top right*), the force moves in a direction opposite to that of the object, and in the second type (*middle right*), the force moves in the same direction as the object. Both levers confer the same mechanical advantage: They gain power at the expense of distance. In the third type (*bottom right*), the force moves over a shorter distance than the object, resulting in a velocity gain. The first two classes of levers have had countless applications in lifting and moving objects and in machinery construction. A detail from an Assyrian bas-relief from Kuyunjik (about 700 B.C.) (*left*) shows a large lever used in moving a giant statue of a man-headed winged bull. *Source*: Bas-relief adapted from Heizer (1966).

smaller forces over longer distances and enlarge the scope of human action (see A4.1).

Levers provide an appreciable mechanical advantage in lifting. Ancient peoples used them extensively for tasks ranging from powering oared ships to moving heavy loads (see Figure 4.1). Nutcrackers and pliers are common hand-tool applications; the wheelbarrow is a less obvious example of a tool that uses leverage. Chinese barrows, used since the Han dynasty, usually had central wheels; with the load right above the axle one of these could carry a huge load. Little sails were sometimes erected to ease propulsion. European barrows, first documented in the high Middle Ages, had their fulcrum at the end; this design put more strain on the person who pushed the barrow than the Chinese design did. The wheel and axle, a circular lever, had an enormous importance for all Old World societies. Invented in Mesopotamia in the late fourth millennium B.C., it diffused rapidly to other parts of the Old World and has seen in countless mechanical uses ever since.

In basic terms, the mechanical advantage of an inclined plane is equal to the quotient of the length of its slope and its vertical height. Friction can reduce this gain quite substantially, and that is why smooth surfaces and some form of lubrication were needed for the best practical performance. According to Herodotus, an inclined plane was the principal means of conveying the heavy stones from the Nile shore to the building site of great pyramids—and there has been much speculation on its further uses during their actual construction. Wedges are just double inclined planes. Many cultures have used wooden wedges to split rocks by inserting them into stone cracks and wetting them. Wedges also form the cutting edges of adzes and axes. Screws are nothing but circular inclined planes. Ancient societies used them to lift water and, much more commonly, in presses to extract plant oils and juices.

A simple pulley offers easy handling of loads but confers no mechanical advantage and carries the danger of accidental weight falls. The ratchet and pawl take care of the last problem, and multiple pulleys solve the first deficiency: The force required to lift an object is nearly inversely proportional to the number of deployed pulleys (see Figure 4.2). Joseph Needham et al. (1965) noted that pulleys were so common in Han China that other things were named after them and even palace entertainments could not do without them. Once a whole *corps de ballet* of 220 dancers in boats was pulled up a slope from a lake! Naturally, good ropes were essential to such feats.

Three classes of mechanical devices—windlasses and capstans, treadwheels, and gearwheels—became critical to lifting, grinding, crushing, and pounding applications. Horizontal windlasses (winches), requiring the shifting of the grip four times a revolution (see Figure 4.3), and vertical capstans (see Figure 4.4) made it possible to transmit power by ropes or chains through simple rotary motion. Cranks—first used in China during the second century and introduced to Europe seven centuries later—made this task even easier.

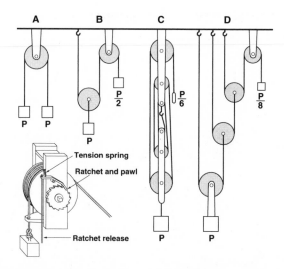

FIGURE 4.2 Equilibrium forces in pulleys are determined by the number of suspension cords. There is no mechanical advantage in *A*. In *B* the weight *P* is suspended by two parallel cords and hence the free end needs to be loaded only by *P/2* to be in equilibrium; in *C* the free end needs to be loaded by *P/6*, and so on. A worker raising building materials with Archimedean potential pulley (*D*) could lift (ignoring friction) a 200-kilogram stone with a force of only 25 kilograms, but a lift of 10 meters would require pulling 80 meters of the counterweight cord. A ratchet and pawl could be used to interrupt this effort anytime.

Windlasses have not only been used to lift water from wells and to raise building materials with cranes but also to wind the most destructive stationary weapons of antiquity, the large catapults that armies used to besiege towns and fortresses (Soedel and Foley 1979). Cranks have been used for numerous purposes, perhaps most notably for powering what became known during the Middle Ages as the Great Wheel, which made it possible to operate lathes with a continuous rotary motion (see Figure 4.5). The arrangement enabled workers to accurately machine the wooden and metal parts used to construct a wide variety of precise mechanisms ranging from clocks to the first steam engines. However, working the wheel while machining hard metals was extremely hard labor. When George Stephenson was manufacturing steel parts for his locomotive in 1813, two strong men found the task so exhausting that they had to rest after every five minutes of work.

The largest muscles of the human body are in the back and legs. Thus, treadwheels delivered much more useful power than hand-turned rotaries. This work could be readily, though not very efficiently, transmitted by gearwheels. Final uses of this motion ranged from lifting loads at construction sites and harbors to operating textile machinery. J. G. Landels (1980), noting that we cannot talk or even think about these machines unemotionally, stressed nevertheless that a well-

FIGURE 4.3 Miners using both a horizontal windlass (*left*) and a crank (*right*) to lift water from a shaft. A heavy wooden wheel, sometimes with pieces of lead fastened to its spokes, helped to conserve the momentum and make the lifting easier. *Source:* Reproduced from Agricola (1556).

designed treadwheel was not only a highly efficient mechanical device but also more comfortable for the operator than a poorly designed one—"in so far as any continuous, monotonous physical work can be comfortable" (pp. 11–12).

Internal vertical treadwheels were a common sight and were especially useful for pumping water (see Figure 4.6). In preindustrial Europe they were also used in large construction and dock cranes. In 1563, Pieter Bruegel the Elder painted such a crane lifting a large stone to the second level of his Tower of Babel (Klein 1978). This device had treadwheels on both sides and was powered by six to eight men. So was one of the last treadwheel cranes (4.8 meters in diameter and 2.7 meters wide) working in the London docks in the nineteenth century.

External vertical wheels were not as common as internal vertical wheels, although they allowed the maximum torque when the operators were treading on a

FIGURE 4.4 Eight men rotating a large vertical capstan in a mid-eighteenth-century French workshop. The capstan winds a cord fastened to pincers, drawing a gold wire through a die. *Source:* Reproduced from Diderot and D'Alembert (1769–1772).

FIGURE 4.5 The great wheel powered by a crank in a mid-eighteenth-century French workshop to turn a metal-working lathe. The smaller wheel was used for working with larger diameters and vice versa. In the background of this picture a man works on a foot-powered lathe machining wood. *Source:* Reproduced from Diderot and D'Alembert (1769–1772).

FIGURE 4.6 Large internal treadwheels were commonly used in raising water—here with a chain-and-ball pump at a German mine—and lifting building materials at construction sites. *Source:* Reproduced from Agricola (1556).

level with the axle. Horizontal treadwheels turning a capstan were fairly popular in Europe. Inclined treadwheels, which laborers operated while leaning against a bar, were also pictured in sixteenth-century sources (see Figure 4.7).

Internal vertical and horizontal treadwheels, as well as inclined wheels, could also be easily adapted for animal operation. All drum-like devices had the added advantage of relatively easy mobility: Operators could take them from job to job by rolling them along on a fairly flat surface. Until the introduction of steam-powered railway cranes, treadwheel-operated cranes remained a highly effective and practical machine for tackling large construction lifting tasks. A treadwheel's maximum power output depended on its size and design. A treadwheel with a single worker delivered no more than 150–200 watts during brief spells of hard effort. The largest treadwheels powered by eight men could deliver as much as 1500 watts.

Animal-operated treadwheels were not always that much more efficient than human-powered ones. The small and often poorly fed beasts of antiquity and the early Middle Ages simply did not perform all that well compared to the draft horses of later centuries. The use of horses for transportation or construction was constrained by the same factors that limited their employment as draft animals in farming. Neither good pastures nor a sufficient supply of feed grains were available in the dry Mediterranean region or in the densely populated lowlands of Asia, and poor harnessing converted their power quite inefficiently. In Asia, domesticated elephants also put a considerable strain on food resources. A classic Indian source of elephant lore extols their effectiveness but prescribes expensive foods for newly caught beasts in training—boiled rice and plantains mixed with milk and sugarcane (Choudhury 1976).

Many other species have been used to perform a variety of tasks around the world. Donkeys, camels, and goats were used for transport and stationary work. In some places dogs turned spits over kitchen fires or pulled small carts or wheelbarrows. Not surprisingly, the modest nutritional demands of bovines—oxen, water buffaloes, and yaks—made them leading working animals both on and off the farm. The typical performance of bovines, however, was at best moderate. For short spells on good roads, they could pull loads as much as three or four times their body weight. When turning a grain mill or powering a water pump, an average bovine could deliver no more than 300 watts.

The old and weak horses that were often used for such work could not deliver much more power than bovines. Stronger seventeenth- and eighteenth-century horses did better when turning a whim, a beam attached to a central axle. This arrangement was commonly used in grain milling or oil pressing. Later it became important for winding in mines (raising coal, ore, or water) and in a variety of industries requiring steady rotary power. Slowly, horses emerged as superior working animals. They powered countless heavy tasks during the early stages of industrial development before the ascent of steam engines. But by that time advanced waterwheels and windmills were being designed.

FIGURE 4.7 Horizontal and inclined treadwheels. The former (*top*), usually equipped with cleats for better traction, were often used to mill grain or to crush ores. Inclined wheels (*bottom*) were less common. *Sources:* Reproduced from Agricola (1556) and Ramelli (1588).

Water and Wind

No longer lay your hand on the millstone, o, ye women who turn the quern. ...
Ceres has commanded her water nymphs to perform the work your arms did, they
fling themselves on the wheel and force round its axle-tree which by means of
mobile rays rotates the mass of four concave mills.

Antipater of Thessalonica (first century b.c.)

People did not begin to harness the energies of flowing water and wind for a surprisingly long time, with the notable exception of ancient sailing ships. Antipater
of Thessalonica, writing during the first century B.C., left the first literary reference to a simple watermill for grain milling. Al-Masudi, writing in 947, left one of
the first reliable records of simple vertical shaft windmills (Forbes 1965). The
Muslim traveler described Seistan (in today's eastern Iran) as the land of winds
and sand where the wind drove the mills and raised the water from streams to irrigate gardens. The successors of these early mills—barely changed from ancient
times, with plaited reed sails behind narrow openings in high mud walls to create
a faster wind flow—could be seen in the region well into the twentieth century.
Both windmills and watermills diffused fairly rapidly throughout the medieval
world, but watermills were far more abundant.

Their ubiquity is attested by the Domesday Book count of 1086, when there
were 5624 mills in southern and eastern England, or one for every 350 people. The
earliest waterwheels were horizontal. They are often referred to as either Greek or
Norse wheels, but the origin of their design remains uncertain. They became
common in many regions of Europe and everywhere east of Syria. The flowing
water, usually directed through a sloping wooden trough on wooden paddles often fitted to a hub at an incline, rotated a sturdy shaft that could be directly attached to a millstone above (see Figure 4.8). This simple but relatively inefficient
design was best suited for small-scale milling.

Vertical wheels, which turned the millstones by right-angle gears, supplanted
the horizontal machines because of their superior efficiency. In Western literature
they became known as Vitruvian mills after the author who first described them
clearly in 27 B.C. Vertical waterwheels are classified according to the point of impact. Undershot wheels were propelled by the kinetic energy of moving water (see
Figure 4.9). These machines had to be located on swift-flowing streams because
their maximum theoretical power was proportional to the cube of the water
speed: Doubling the speed boosted the capacity eightfold.

Where the stream flow was first impounded, undershots were used only with
heads (differences in water level) of between 0.5 and 3 meters. The Poncelet wheel,
with curved blades, introduced around 1800, was the most efficient undershot
machine. Wheel diameters were roughly three times as large as the head for paddle wheels in general, and up to four times as large for Poncelets. These low-head,
slowly rotating machines converted about 20 percent of water's kinetic energy

FIGURE 4.8 The horizontal waterwheel, also called a Greek or Norse wheel, was powered by the impact of running water and rotated the runner stone directly. *Source:* Reproduced from Ramelli (1588).

FIGURE 4.9 Engravings of a large undershot wheel running a French royal paper mill (*top*) and of an overshot wheel (*bottom*) powering an ore-washing machinery in a French forge. *Source:* Reproduced from Diderot and D'Alembert (1769–1772).

into useful power; by contrast, nineteenth-century wheels could deliver up to 35 to 45 percent.

Breast wheels were turned by a combination of water flow and gravity in streams with heads of between 2 and 5 meters. Close-fitting breast works prevented premature water spillage and were essential for good performance. In low-breast designs, the water entered below the elevation of the center shaft; these machines were no more efficient than well-designed undershots. In high-breast machines, the water struck the wheel above the elevation of the center shaft; these wheels approached the outputs of overshot wheels.

Traditional overshots operated with heads of more than 3 meters, and their diameters were usually equal to about three-quarters of the head. The water was fed through troughs or flumes into bucket-like compartments at rates ranging from less than 100 to more than 1000 liters per second. Most of the rotary power (common speeds were 4 to 12 revolutions per minute) was generated by the weight of descending water (see A4.2). Consequently, overshots could be placed on streams with slow water flow. This advantage was partially negated by the need for a well-directed and carefully regulated water supply, which required storage ponds and races to be built and maintained. Overshots operating with excess carrying capacity, that is, with reduced spillage from buckets, could be more efficient, although naturally less powerful, than machines under full flow. Top overshot efficiencies were high: They could convert up to 85 percent of water's potential energy into rotary motion, and performances at around 60 percent were the norm.

Undershots could be placed directly in a stream, but such a location naturally increased the chances of flood damage. Breast wheels and overshots needed a regulated water supply. Usually, a weir constructed across a stream regulated the water flow and a channel diverted the flow to the wheel. In regions of low or irregular rainfall it was common to impound water in ponds or behind low dams. No less attention had to be paid to returning water to the stream because backed-up water would have impeded wheel rotation. Smooth tail races were also needed to prevent channel silting. Even in England, wheels, shafts, and gears were almost invariably made of wood until the beginning of the eighteenth century. Afterwards came a growing use of cast iron for hubs and shafts. The first all-iron wheel was built early in the nineteenth century (Crossley 1990).

In addition to fixed stream waterwheels, there were also the much less common floating wheels, which could be installed on barges and tidal mills. Floating grain mills were successfully used for the first time on the Tiber; in 537, the Goths besieged Rome and cut off the aqueduct water turning the milling wheels. Floating mills were not an infrequent sight on the rivers in or near the cities and towns of medieval Europe, and many remained until the eighteenth century. The use of intermittent power from the sea was first documented in Basra during the tenth century. During the Middle Ages, small tidal mills were built in England, in the Netherlands, on the Atlantic coast of the Iberian peninsula and, above all, in Brittany. Later came the installations in North America and in the Caribbean

(Minchinton and Meigs 1980). Most of these mills had vertical wheels working only with the ebbing tide. Larger mills had reservoirs extending the operation to 16 hours a day.

Gradually, waterwheel uses expanded far beyond grain milling. Waterwheels came to power scores of previously manual tasks. A brief list must include sawing, wood turning, oil pressing, paper making, cloth fulling, tanning, ore crushing, iron making, wire pulling, stamping, cutting, metal grinding, blacksmithing, wood and metal burning, majolica glazing, and polishing. Waterwheels performed all these tasks with much higher levels of efficiency and productivity than manual labor could manage.

Moreover, the unprecedented magnitude, continuity, and reliability of power provided by waterwheels opened up new productive possibilities. This was especially true in mining and metallurgy. Indeed, to a significant degree, the energy foundations of Western industrialization rest on these specialized uses of waterwheels. Human and animal muscles could never generate power at such high, concentrated, continuous, and reliable rates—but only such deliveries could increase the scale, speed, and quality of countless food-processing and industrial tasks. Yet it took a long time for typical waterwheels to reach capacities surpassing the power of large harnessed animal teams. For centuries, the only way to achieve larger power outputs was to install a series of smaller units in a suitable location. A famous Roman mill-line at Barbegal near Arles had 16 wheels, each with about 2 kilowatts of capacity, for a total of just over 30 kilowatts (Sellin 1983).

Such installations remained rare for centuries to come. Even during the early decades of the eighteenth century European waterwheels averaged outputs of less than 4 kilowatts. Only a few machines surpassed 7 kilowatts, and crude finishing and poor gearing resulted in low conversion efficiencies. Even the most admired machines of that time—fourteen large waterwheels (12 meters in diameter) built on the Seine at Marly between 1681 and 1685 in order to pump water for the Versailles fountains of Louis XIV—were capable of less than 7 kilowatts. Their actual useful output was just over 4 kilowatts (Klemm 1964).

But when combined with innovative machine design, the late eighteenth-century waterwheels made an enormous difference in productivity. For example, a water-powered machine for cutting and heading 200,000 nails a day received a U.S. patent in 1795 (Rosenberg 1975). Widespread adoption of these machines brought nail prices down by nearly 90 percent during the next fifty years. The concentrated power needs of late eighteenth-century industrialization forced the pace of capacity growth, both in unit and aggregate terms. By 1840, the largest British installation—Shaw's waterworks at Greenock, near Glasgow on the Clyde—had thirty mills built in two rows on a steep slope. They were fed from a large reservoir to provide about 1.5 megawatts of power. The largest individual waterwheels had diameters around 20 meters, widths of 4 to 6 meters, and capacities well above 50 kilowatts.

The world's largest wheel was the Lady Isabella. A pitchback overshot machine with a diameter of 21.9 meters, it was built in 1854 by the Great Laxey Mining Company on the Isle of Man (Reynolds 1970). All the streams on the slope above the wheel were channeled into the collecting tanks, and water was then piped into the base of the masonry tower and into a wooden flume. The power was transmitted to the pump rod, which reached 451 meters to the bottom of the lead-zinc mine shaft, by the main-axle crank and by 180 meters of timber connecting rods. The wheel's theoretical peak power was about 427 kilowatts. In normal operation it generated about 200 kilowatts of useful power.

But the era of giant waterwheels was short-lived. Just as these machines were being built during the first half of the nineteenth century, water turbines were being developed. These machines brought the first radical improvement of water-driven prime movers since the introduction of vertical wheels centuries before. Benoit Fourneyron built his first reaction turbine in 1832 to power forge hammers. It featured a radial outward flow, a very low head of just 1.3 meters, and a rotor diameter of 2.4 meters; even so, it had a capacity of 38 kilowatts. Five years later two improved Fourneyron machines working at a Saint Blaisien spinning mill rated at about 45 kilowatts (Smith 1980)

Fourneyron's machines were soon surpassed by James B. Francis's innovative inward-flow turbines (invented in 1847). Later came the jet-driven turbines of Lester A. Pelton (patented in 1889) and the axial flow turbines of Viktor Kaplan (in 1920). New turbine designs replaced waterwheels as the prime movers in many industries. For example, in Massachusetts they accounted for four-fifths of the installed power by 1875. By then, water-driven machines had reached the peak of their importance for the rapidly industrializing society.

Each of the three leading textile mill centers on the lower Merrimack in Massachusetts—Lowell, Lawrence, and Manchester—had water machines; the combined power totaled about 7.2 megawatts. The whole river basin had about 60 megawatts of installed capacity; this amount averaged out to some 66 kilowatts per manufacturing establishment (Hunter 1975). Even in New England in the mid-1850s, steam was still about three times more expensive than water as a prime mover. But the era of water turbines ended rather abruptly. By 1880 large-scale coal mining and more efficient engines had made steam cheaper than water power virtually everywhere in the United States. Before the end of the nineteenth century most water turbines were turning electricity generators instead of delivering direct power.

In a number of Asian and European regions, windmills were the most powerful prime movers of the preindustrial era. Naturally, they were especially important in dry regions with seasonally strong winds (the Middle East, Greece, and Portugal, for example) and in flatlands where waterwheels were not feasible because water heads were almost nonexistent (in the Netherlands, Denmark, and parts of England). But their overall contribution to worldwide economic intensification was less decisive than that of waterwheels.

In Europe they appeared rather suddenly (the first clear records come from the last decades of the twelfth century) and spread fairly rapidly. Unlike the simple Eastern machines with horizontally mounted sails, European windmills were vertically mounted rotaries with drive shafts that could be turned into the wind. With the exception of Iberian octagonal mills, which had triangular sails—a clear import from the eastern Mediterranean—early European machines were all post mills. The wooden structure housing the gears and millstones pivoted on a massive central post that was usually supported by four diagonal quarterbars (see Figure 4.10). The whole engine-house had to be turned into the wind. Post mills were fragile structures requiring constant attention. They were unstable in high winds and vulnerable to storm damage, and their relatively low height limited their performance (see A4.3).

Although post mills continued to work in parts of Eastern Europe until the twentieth century, in the West they were gradually replaced by tower mills and smock mills. In both of these designs, only the top cap was turned into the wind, either from the ground or from galleries (see Figure 4.11). Smock mills had wooden frames, usually octagonal, covered with clapboards or shingles. Tower mills were typically rounded and tapered stone structures. The English fantail, a device that powered a winding gear that turned the sails automatically into the wind, was not introduced until after 1745. Curiously, the Dutch, who operated the largest number of windmills in Europe, adopted this innovation only in the early nineteenth century.

But Dutch millers were the first to introduce more efficient blade designs. They added a canted leading edge to previously flat blades in around 1600. The resulting arch (camber) gave blades more lift while reducing drag. Later innovations included improvements in sail mounting, cast metal gearings, and centrifugal regulating governors. These devices did away with the difficult and often dangerous task of adjusting the canvas to different wind speeds. By the end of the nineteenth century the English had started to install true airfoils, aerodynamically contoured blades with thick leading edges.

Windmills were usually used for milling grain and pumping water. The latter could also be performed on ships with small portable machines. They were also used both in Europe and in the Islamic world for grinding and crushing a variety of substances (chalk, sugar cane, mustard, cocoa), making paper, sawing, and working metal (Hill 1984). In the Netherlands they did all of these things, but their greatest contribution was in draining the country's low-lying land and reclaiming polders for cropfields. The first Dutch drainage mills date from after 1300, but they became common only in the sixteenth century.

America's westward expansion on the windy Great Plains created a need for smaller, simpler, yet efficient machines to serve railway stations and farms. Instead of a few large, wide sails, American windmills usually had a large number of fairly narrow blades or slats that were fastened to solid or sectional wheels. They were commonly equipped with either a centrifugal or a side-vane governor and

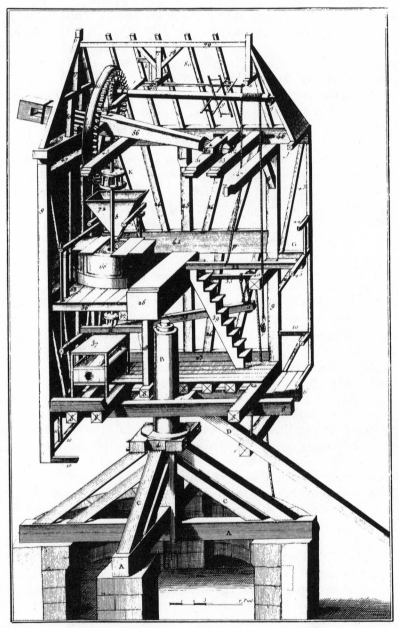

FIGURE 4.10 A French wooden post mill. The main post, almost always oaken, was held up by four quarterbars attached to massive cross-trees. Windmill rotations were transferred to the millstone by a lantern and crown gear. The only access was by ladder.
Source: Reproduced from Diderot and D'Alembert (1769–1772).

FIGURE 4.11 A detailed modern drawing of a traditional English smock windmill. Herne mill in Kent was built in 1789 and later raised onto a two-story brick base. The tapering tower housed the gears and millstones. The sail radius was about 10.5 meters, and the vane attached to the mill's top kept the sails turned into the wind. *Source:* Reynolds (1970). Reprinted by permission of the publisher.

with independent rudders, placed on top of lattice towers. These windmills were used mostly to pump water for households and cattle and for steam locomotives, an indispensable application during the westward extension of railways (see Figure 4.12).

Windmills reached the peak of their importance during the nineteenth century. The United Provinces of Netherlands already had at least 8000 machines in 1650, but some 10,000 mills were working in early nineteenth-century England, and a nationwide German count reached more than 18,000 units in 1895. In 1900, about 30,000 mills with a total capacity of some 100 megawatts were working in countries around the North Sea (DeZeeuw 1978). In the United States, several million Halladay, Adams, and Buchanan windmills and scores of the smaller brands were sold during the second half of the nineteenth century. Large numbers of American water-pumping windmills were used during the twentieth century in Australia, South Africa, and Argentina.

The capacities of early windmills are unknown. The first reliable experimental measurements date from 1760, when John Smeaton concluded that the power of a common Dutch mill with 9-meter sails was equivalent to the power of ten men or two horses (Smeaton 1760). This calculation, based on measurements with a small model, was corroborated by actual performance in oilseed pressing. The windmill-powered runners turned 7 times a minute, but two horses made scarcely 3.5 turns in the same time.

According to Robert Forbes (1958), a typical large eighteenth-century Dutch mill with a span of 30 meters had a capacity of about 10 horsepower (7.5 kilowatts). Modern measurements at a well-preserved mid-seventeenth-century Dutch drainage mill indicated a windshaft power of about 30 kilowatts, but large transmission losses lowered the useful output to less than 12 kilowatts. All of these results confirmed William Rankine's comparison of traditional prime movers. He credited post windmills with 1.5 to 6 kilowatts of useful power and tower mills with 4.5 to 10.5 kilowatts (Rankine 1866). Similar measurements of American windmills put their useful power across a wide range, from a mere 30 watts for the 2.5-meter wheels to as much as 1 kilowatt for the large 7.6-meter machines (Wolff 1900).

The capacities of common post windmills and watermills during medieval times were about the same (1 to 2 kilowatts). But by the early nineteenth century there was a huge gap between these mechanical prime movers. Many waterwheels were four to five times more powerful than even the largest tower mills, delivering 8 to 12 kilowatts of useful power. This difference only grew with the subsequent development of water turbines.

Explosives

Gunpowder is pressed to fill six-tenths of the barrel; then two-tenths of the barrel is filled with fine earth which is packed very gently. Then two or three pint

FIGURE 4.12 The Halladay windmill. During the last decade of the nineteenth century these windmills were the most popular American brand. They were a common sight in Western railway stations where they pumped water for steam locomotives. *Source:* Reproduced from Wolff (1900).

measures of iron balls are put in. ... The shots go with a force that destroys things along their path like the breaking of dried twigs.

—Huo Lung Qing (1412)

Any account of preindustrial prime movers would be incomplete without at least a brief survey of the origins, diffusion, and historical importance of gunpowder. All early and classical warfare was strictly powered by animate prime movers. Warriors wielded daggers, axes, and swords in close combat. They used spears and lances, drew bows and much more powerful crossbows (both the Chinese and the Greeks used such weapons beginning in the fourth century B.C.), wound the winches of massive catapults, and handled odd-looking siege machines. Horses, camels, and elephants transported supplies, pulled chariots, or carried charging men on their backs. After the general adoption of stirrups, horses and men created an especially effective fighting unit. Asian riders—unarmored but with extraordinarily hardy horses—were distinguished by their speed and maneuverability. European knights—heavily armored astride large animals—relied on their relatively high invulnerability.

Every old high culture held thunder and lightning in awe. The aspiration to emulate their destructive power recurred in many narratives and myths (Lindsay 1974). For millennia warriors attached incendiary materials to arrowheads or hurled such materials from catapults. Sulfur, petroleum, asphalt, and quicklime were used in these incendiary mixtures. But only gunpowder combined propulsive force with great explosive and inflammatory power.

Gunpowder's origins undoubtedly stem from the long experience of Chinese alchemists and metallurgists (Needham et al. 1986). They were familiar with the three ingredients—potassium nitrate (KNO_3, saltpeter), sulfur, and charcoal—long before they started to combine them. The first incipient gunpowder formula comes from a Taoist classic of the mid-ninth century. Clear directions for preparing gunpowders for different kinds of bombs were published in 1040. The early mixtures were not truly explosive because they contained only about 50 percent saltpeter.

Eventually the proportions settled to around 75 percent saltpeter, 15 percent charcoal, and 10 percent sulfur, the mixture capable of detonation. Unlike ordinary combustion, where oxygen must be drawn from the surrounding air, ignited KNO_3 readily provides its own oxygen; gunpowder rapidly produces a roughly 3000-fold expansion of its volume in gas. Obviously, when appropriately confined and directed, such a force can be used to propel heavy projectiles. After the invention of gunpowder, cannons and firearms followed rather rapidly.

Artillery developments started with the Chinese fire-lances of the tenth century. These tubes that ejected bits of material were first made of bamboo and later of metal; they evolved into simple bronze cannons that could hurl stones. The first true guns were cast in China before the end of the thirteenth century and in Europe in the early decades of the fourteenth century. The pressures of frequent armed conflicts led to rapid rates of innovation resulting in more powerful and more accurate guns. Their power increased with the introduction of iron cannon-balls during the fifteenth century.

The strategic implications of gunpowder warfare were immense, both on the land and on the seas. There was no need to mount prolonged and often desperate sieges of seemingly impregnable castles. The combination of accurate artillery and iron cannonballs, which were much more destructive than their stone prede-cessors because of their higher density, made the castles indefensible. Attackers able to destroy sturdy stone structures from far beyond the range of archers put an end to the defensive value of traditionally built castles and walled cities.

The relatively compact medieval fortresses with thick stone walls were super-seded by new designs. The most common were the low spreading star-shaped polygons with massive earthen embankments and huge water ditches. These proj-ects consumed enormous amounts of materials and energy. The fortifications of Longwy (in northeastern France), the largest project of the famous French mili-tary engineer Sebastien Vauban (1633–1707), required 640,000 cubic meters of rock and earth and 120,000 cubic meters of masonry (M. S. Anderson 1988).

On the seas during the era of aggressive expansion, gunned ships were the prin-cipal carriers of European technical supremacy to distant locations. Long-range guns gave the English captains the decisive advantage over the Spanish Armada in 1588. A century later, large men-of-war were fitted with up to 100 guns, and the British and Dutch ships engaged during the battle of La Hogue in 1692 carried a total of 6756 guns (M. S. Anderson 1988). The weaponry of this era reached levels that were not decisively surpassed until the middle of the nineteenth century with the introduction of nitrocellulose-based powders (during the 1860s) and dyna-mite (patented by Alfred Nobel in 1867).

Biomass Energies

There was a time when the trees were luxuriant on the Ox Mountain. As it is on the outskirts of a great metropolis, the trees are constantly lopped by axes. Is it any wonder that they are no longer fine?

—Meng Ke (Mencius) (fourth century b.c.)

Virtually all heating, lighting, and metallurgical needs in traditional societies were provided by wood, charcoal, crop residues, dried dung, and plant and ani-mal oils and waxes. Acquiring these fuels could be as easy as making a short trip to a nearby forest, bush, or field to collect fallen branches, twigs, straw, or dry grasses. But it could also entail laborious tasks like cutting down trees, making charcoal, and transporting materials over long distances in ox-drawn carts or in camel caravans. Fuel availability shaped house design, clothing, and cooking practices. Because wood supplied most of the fuel in traditional societies, defores-tation was not uncommon.

Although few quantitative details are known about the consumption of plant fuels in preindustrial societies, a large number of recent studies examine biomass energy use in traditional cultures (Earl 1973; Openshaw 1978; Hall et al. 1982; Smil 1983). These studies help to fill in our gaps of information because the basic needs have not changed. In most cultures families had to cook two or three meals a day; in cold climates they had to heat at least one room; and in some societies they also had to prepare feed for animals and dry some foods for storage.

People have used wood in any available form, including fallen, broken, or lopped-off branches, twigs, bark, and roots—but used chopped stemwood only where good cutting tools (adzes, axes, and later, saws) were commonly available. The variety of the wood made surprisingly little difference. There are thousands of woody plants, but their chemical composition is remarkably uniform. They are roughly two-fifths cellulose, one-third hemicellulose, and lignin. But the physical differences are substantial: The specific mass of some oaks, for example, is almost twice as high as that of some poplars.

The energy content of wood rises with the proportion of lignin and resins, but the differences among common woody species are fairly small (see A4.4). Standard energy densities always refer to absolutely dry matter, but wood burned in traditional societies had a widely varying moisture content. Freshly cut mature hardwoods (leafy trees) are typically 30 percent water, whereas softwoods (conifers) are well over 40 percent water. Softwood burns inefficiently because a significant part of the released heat goes into vaporizing moisture rather than heating a cooking pot or a room.

When the moisture content of wood is more than 67 percent, the fuel will not ignite (Tillman 1978). That is why dry fallen branches and twigs or hacked-off pieces of dead trees have always been preferable to fresh wood and why wood is usually air-dried before combustion. In traditional societies, cut wood was stacked, sheltered, and left to dry for at least a few months, but even in dry climates it still retained about 15 percent moisture. In contrast, charcoal contains only a trace of moisture and has always been preferred by those who could afford it.

Charcoal is virtually smokeless, and its energy content, equal to that of good bituminous coal, is roughly 50 percent higher than that of air-dried wood. Charcoal's other main advantage is its high purity. Because it is virtually pure carbon, it contains hardly any sulfur or phosphorus. This feature makes it the best possible fuel not only for indoor use but also for producing bricks, tiles, and lime in kilns and for smelting ores. It has a further advantage for smelting because its high porosity (a specific density of merely 0.13 to 0.20, compared to 2.0 for soot) facilitates the ascent of reducing gases in furnaces (Sexton 1897).

But traditional production of this excellent fuel was very wasteful. Partial combustion of the heaped wood inside primitive earth or pit kilns generated the heat necessary for carbonization. Consequently, there was no need for additional fuel, but both the quality and the quantity of the final product was difficult to control (see Figure 4.13). Typical charcoal yields in such kilns were only between 15 and 25 percent of the air-dried wood. This means that about 60 percent of the original energy was lost in making charcoal.

Crop residues were indispensable fuels for societies living on deforested, densely settled agricultural plains or in arid, sparsely treed regions. Cereal straws and stalks were usually the most abundant residues, but many other types were locally and regionally important. These included legume straws and tuber vines, cotton stalks and roots, jute sticks, sugar cane leaves, and branches and twigs pruned from fruit trees. Some crop residues required drying before combustion.

Ripe straws are only between 7 and 15 percent water, and their energy content is comparable to that of hardwood. But their density is obviously much lower than that of hardwood; thus, storing enough straw to last through winter could never be as easy as stacking chopped wood. The low density of crop residues also meant that fires and stoves had to be stoked almost constantly. Because of a number of competitive uses, crop residues were often in short supply. Straw makes a good food for ruminants, serves well for animal bedding and roof thatch, and can be used in the manufacture of simple tools and domestic articles. Legume residues were an excellent high-protein feed and fertilizer.

Consequently, every bit of combustible biomass was often gathered for household use. Middle Eastern peoples burned thorny shrubs and substituted date kernels for charcoal. On the North China Plain women and children collected fallen twigs, leaves, and dry grasses with rakes, sickles, baskets, and bags (King 1927). And in the interior of Asia, throughout the Indian subcontinent, and in parts of the Middle East, Africa, and both Americas, dried dung was the most important source of heat for cooking. The heat value of air-dried dung is comparable to that of crop residues or grasses (see A4.4).

Dung has made important contributions to many societies; in fact, it was essential to America's westward expansion (Welsch 1980). Wild buffalo and cattle dung made possible the early continental crossings and the subsequent colonization of the Great Plains during the nineteenth century. Travelers on the Oregon and Mormon Trails collected "buffalo wood," and the early settlers stacked winter supplies in igloo shapes or against house walls. Known by many euphemisms (cow wood, Nebraska oak), the fuel burned evenly and with little smoke or odor. However, it burned quickly and required almost continuous stoking.

In South America, llama dung was the principal fuel on the Altiplano of the Andes, the core of Inca empire in today's southern Peru, eastern Bolivia, and northern Chile and Argentina (Winterhalder et al. 1974). Cattle and camel dung was always used in the Sahelian region of Africa as well as in many Egyptian villages. Cattle dung was gathered in large quantities in arid as well as in monsoonal Asia, and Tibetans relied on yak dung. Only sheep dung has been generally avoided because it produces an acrid smoke when burned. In India, where dung use is still widespread, both cow and water buffalo droppings are gathered daily, mostly by *harijan* (untouchable) children and women, both for their own house-

FIGURE 4.13 Charcoal-making procedures were very similar throughout the Old World. Both pictures—the top one from late seventeenth-century England, the bottom from mid-eighteenth-century France—show workers leveling the ground, setting up the central pole, stacking the cut wood, and covering up the finished pile before ignition. *Sources:* Reproduced from a 1679 illustration in Evelyn (1706) and from Diderot and D'Alembert (1769–1772).

hold use and for sale (Patwardhan 1973). They are collected either as dry chips or as a fresh biomass. Fresh dung is mixed with straw or chaff, hand-molded into patties and cakes, and sun-dried in often ornamentally adorned piles or small pyramids.

Studies of traditional societies that remained dependent on biomass fuels into the second half of the twentieth century indicate annual fuel requirements of less than 500 kilograms per capita in the poorest villages of tropical regions. Pronounced winters and substantial wood-based production of building materials and metals could increase this need five times over. Just before their switch to coal, western Europe and North America used even greater amounts.

Some nineteenth-century northern European, New England, Midwestern, and Canadian communities heating and cooking only with wood annually consumed anywhere from 3 to 6 tonnes of fuel per capita. Indeed, even America's nationwide average in the middle of the nineteenth century was about 5 tonnes per capita (Schurr and Netschert 1960). Although that figure included growing industrial (mainly metallurgical) and transportation uses, household combustion was still the leading consumer of American wood during the 1850s.

Household Needs

The things that people cannot do without every day are firewood, rice, oil, salt, sauce, vinegar, and tea.

—Chinese proverb

Every settled society had to process some crops before they could be eaten directly or used in the preparation of more elaborate foods. Grain milling has been virtually universal, and historically, it was almost always the first processing step that a society developed. Later came extraction of oils by pressing a variety of seeds, fruits, and nuts. Many societies also processed tubers in order to remove antinutritive factors or prepare them for storage. Some crushed sugar cane to express its sweet juice. In all these tasks, human energy was only gradually augmented by animal labor. The first use of machinery to mill grains occurred about 2000 years ago when horizontal waterwheels were adapted to the task.

Cooking required relatively little energy in East Asian stir-frying and steaming. In contrast, considerable fuel inputs were needed for baking bread throughout the rest of the Old World and for the roasting commonly practiced in the Middle East, Europe, and Africa. In some societies fuel was also required to prepare feed for domestic animals, above all for pigs. Seasonal heating of dwellings was necessary in all mid-latitude and boreal cultures, but preindustrial houses were generally heated for surprisingly short periods and to relatively low indoor temperatures.

In some fuel-short regions there was no winter heating at all in spite of months of low temperatures. For example, in the deforested lowlands of Ming and Qing China there was no winter heating south of the Yangzi River, although the northernmost parts of this region have mean January and February temperatures between just 2 and 4 °C, with minima going below –10 °C (Domros and Peng 1988). And the chill of traditional English interiors, even after the introduction of coal stoves, has been proverbial.

The total household energy needs of East Asian or Middle Eastern societies were thus very low. The absolute fuel demand of some northern European and colonial North American communities was rather high, but these figures can be misleading because the combustion efficiencies were often so low. High use could result in relatively low shares of useful heat. Consequently, even in nineteenth-century America, endowed with plenty of fuelwood, an average household claimed only a small fraction of the useful energy flows available to a twentieth-century household.

Food Preparation

In forest lands we hack at firewood,
with kindling for the fires to cook our food.
—Shi jing *(Book of Poetry)* (eleventh century b.c.)

Given the dominance of cereals in the nutrition of all high cultures, the milling of grains was certainly the most important food-processing technique (Bennett and Elton 1898). Whole, unprocessed grain is not very palatable, it is difficult to digest, and obviously, it cannot be used for baking. Milling produces flours of various fineness that can be used to make highly digestible foods—above all, breads and noodles.

The first tools for milling grain were slightly hollowed rubbing stones and stone pestles and mortars (see Figure 4.14). Ancient Middle Eastern societies and the peoples of preclassical Europe used an oblong, oval saddlestone that could be worked from a kneeling position. Push mills with hoppers and grooved bedstones were the first major innovation. The Greek hourglass mill had a cone-shaped hopper and a conical grinder. Manual rotary querns appeared just before the Christian era. The productivity of all muscle-driven processing was very low (Moritz 1958). The tedious manual grinding yielded no more than 2 or 3 kilograms of roughly ground flour per hour. In Greece and Rome, two men, usually slaves, turning hand querns worked at a rate of no more than 200 watts. They could grind only 7 kilograms of grain in one hour. A donkey powering an hourglass mill with about 300 watts could grind around 12 kilograms of flour per hour.

Although antique and early medieval waterwheels had limited power, they

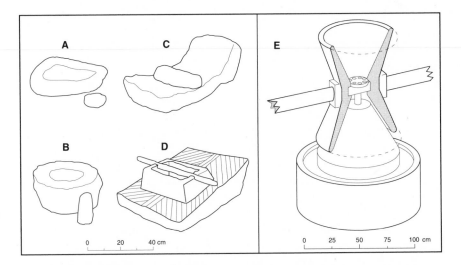

FIGURE 4.14 Grain-milling tools. Primitive millers first used simple rubbing stones (*A*) and then progressed to stone pestles and mortars (*B*). Saddlestones (*C*) were first used in Egypt just before the beginning of the Old Kingdom. The Greek push mill (*D*), dating from around 500 B.C., had a grooved surface for better grinding. The hourglass mill (*E*), used both by the Greeks and the Romans, was powered either by slaves or donkeys. *Source:* Based largely on Fox (1976).

were much more efficient than hand-turned or animal-powered stones. Even the simplest vertical wheels were equivalent to three horses turning a whim. A millstone driven by a waterwheel with a capacity of 1500 watts produced between 80 and 100 kilograms of flour in one hour. It is reasonable to assume that bread supplied about half of the energy in an average diet. Consequently, a single mill would have produced enough flour in a ten-hour shift to feed 2500 to 3000 people, a fair-sized town, for one day.

Horizontal waterwheels could rotate millstones directly, but vertical waterwheels, and all windmills, transmitted their rotary power by means of wooden gears. And no mill could produce good flour without accurately set and well-dressed millstones (Freese 1957). By the eighteenth century the two stones—the top runner and the stationary bedstone—were usually 1–1.5 meters in diameter, up to 30 centimeters thick, weighed close to 1 tonne, and rotated 125–150 times a minute. Grain was fed from the hopper into the opening (eye) of the runner then crushed and milled between lands, the stones' flat surfaces.

These massive stones had to be balanced with great precision. If they rubbed against each other they could be badly damaged, and they could also easily spark a fire. If they were too far apart they produced rough meal rather than fine flour. The tolerances were small: No more than the thickness of heavy brown paper could separate the stones at the eye—and no more than a tissue-paper thickness at the edge!

Ground flour and milling by-products were channeled outwards along incised grooves (furrows). Skilled craftsmen used sharp tools (mill bills) to deepen these furrows (dress the stone). They did this at regular intervals determined by the quality of the stone and the rate of milling, usually every two to three weeks. Many kinds of stone were used, but solid granites or hard sandstones, or pieces of cellular quartz (buhrstones) cemented together and held by iron hoops, were the most common choices. No millstone could do a perfect job in a single run. After the separation of the coarse bran from the fine flour, the miller returned the intermediate particles to the stones to be reground. The whole process could be repeated several times (Snyder 1930). The final step, sieving (bolting), separated the flour from the bran and sorted the flour into different grades.

For centuries, milling grain still required a great deal of heavy labor, even with windmills and waterwheels. Laborers had to unload the grain, hoist it with pulleys into hoppers, cool the freshly milled flour by raking it, sort it with sieves, and load it into bags. Sieves driven by water power were introduced during the sixteenth century. An American engineer, Oliver Evans, designed a fully automatic flour mill in 1785. His invention used endless bucketed belts to lift the grain and augers (Archimedean screws) to transport it horizontally and to spread the freshly ground flour for cooling. Although the mill was not an immediate commercial success, his self-published book on milling became a classic of the genre (Evans 1795).

There were remarkably few advances in cooking methods until the onset of the industrial era. People of old high cultures used open hearths and fireplaces for roasting (in the fire or on spits, skewers, or gridirons), boiling, frying, and stewing. They used braziers for boiling water and for grilling, and simple clay or stone ovens for baking. Flat breads were stuck to the sides of clay ovens, and leavened breads were placed on flat surfaces. Fuel shortages contributed to the popularity of low-energy cooking methods. The Chinese were already using cooking pots that stood on three hollow legs (*li*) before 1500 B.C. Shallow sloping pans—the Indian and Southeast Asian *kuali* and the Chinese *kuo*, better known in the West as the Cantonese *wok*—speeded up frying, stewing, and steaming (E. N. Anderson 1988).

The origin of the kitchen stove remains uncertain, but its wide acceptance obviously necessitated the construction of chimneys. Even in the richest parts of Europe, chimneys were uncommon before the beginning of the fifteenth century. The Chinese used clay or brick stoves, but many homes still did not have chimneys during the first decades of the twentieth century (Hommel 1937). Iron stoves fully enclosing the fire started to replace open fireplaces for cooking and heating only during the eighteenth century. Benjamin Franklin's famous stove, conceived in 1740, had its European counterpart in the Rumford roaster introduced in 1798.

Heat and Light

I suppose, with others, that cold is nothing more than the absence of heat or fire.
—Benjamin Franklin, in a letter of April 14, 1757

Traditional methods of providing heat and light were often primitive and ineffi-cient—an especially remarkable fact given the often impressive mechanical inven-tions of ancient civilizations. For millennia of preindustrial evolution, people heated their dwellings with open fires and simple fireplaces. The glow of the fire and the flickering, weak flames of oil lamps and candles provided poor illumina-tion.

In heating, it took a long time for traditional peoples to make the transition from wasteful, unregulated open fires to more efficient arrangements. First, they merely moved the open fire into a three-sided fireplace, but this step brought only a marginal gain in efficiency. Well-stoked fireplaces could keep unattended fires burning overnight but were not very efficient. The best efficiency rates were close to 10 percent, but typical performances averaged around 5 percent. And often a working fireplace warmed its immediate vicinity but drew in cold air from out-side; in this way, it could actually cause an overall heat loss in the room.

The efficiencies of traditional brick or clay stoves varied—not only with design (often mandated by cooking preferences) but also with the type of fuel. Modern measurements of Asian rural stoves, which have not changed for centuries, make it possible to fix the highest practical efficiencies. Grated, massive brick stoves fu-eled by chopped wood (and built with long flues and tightly fitting tops) were about 20 percent efficient; less massive, drafty stoves fueled with straw or grasses (with short flues) were just 15 or 10 percent efficient.

But not all traditional heating arrangements were wasteful. At least three space heating systems used wood and crop residues in ingeniously efficient ways while providing a great degree of comfort. They were the Roman *hypocaust* (introduced around 80 B.C.,) the Korean *ondol*, and the Chinese *kang*. The first two designs led hot combustion gases through raised floors before exhausting them through a chimney. Trials with a preserved hypocaust showed that no more than 1 kilogram of charcoal was needed per hour to maintain a very comfortable temperature—22 °C—in a room of $5 \times 4 \times 3$ meters even when the outside temperature was at the point of freezing (Forbes 1966). The kang is still found throughout North China. In this system, the waste heat from the kitchen stove warms a large brick platform (at least 2 by 2 meters and 75 centimeters tall). The platform serves as a bed at night and as a resting place during the day (Hommel 1937).

These systems conducted heat slowly over relatively large areas. In contrast, brazier heaters, common in most Old World societies, could heat only small areas and produced high concentrations of carbon monoxide. The Japanese, although great exploiters of Chinese and Korean inventions, could not introduce the ondol or the kang into their flimsy wooden houses. They relied instead on charcoal braziers (*hibachi*) and on footwarmers (*kotatsu*). These small containers of charcoal, which were set into the floor and covered with wadded cloth, were used well into the twentieth century. They survive even today: The electric kotatsu is a small heater built into a low table. Charcoal heaters were not uncommon elsewhere— even the British House of Commons was heated by large charcoal fire pots until 1791.

Biomass fuels also provided most of the lighting in preindustrial societies. Fires, torches made of resinous wood, and burning splinters were the simplest but also the least efficient and least convenient methods of providing light. Fat-burning oil lamps appeared in Europe during the Upper Paleolithic nearly 40,000 years ago (de Beaune and White 1993). Candles were used in the Middle East only after 800 B.C. Both oil lamps and candles offered inefficient, weak, and smoky illumination, but they were at least easily portable and safer to use. They burnt a variety of animal and plant fats and waxes—olive, castor, rapeseed, linseed and whale oil, beef tallow, and beeswax—with papyrus, rush pith, flaxen, or hempen wicks. Until the end of the eighteenth century, artificial indoor light came only in units of one candle. Bright illumination was possible only through massive multiplication of these tiny sources.

Candles convert only about 0.01 percent of their chemical energy into light. The bright spot in a candle flame has an average irradiance (rate of energy falling on a unit area) just 20 percent higher than clear sky. The first eighteenth-century lighting innovations doubled and tripled this low performance. In 1794, Aime Argand introduced lamps that could be regulated for maximum luminosity by using wick holders with central air supply and chimneys to draw in the air (McCloy 1952). Soon afterward came the first lighting gas made from coal. For more than half of the nineteenth century, tens of millions of rural households around the world continued to fill their lamps with an exotic biomass fuel—the oil rendered from the blubber of sperm whales.

Whaling, a poorly paid, wearying, and dangerous way of life—portrayed so unforgettably in Melville's (1851) great book—reached its peak just before 1850 (Francis 1990). The American whaling fleet, by far the largest in the world, had a record total of more than 700 vessels in 1846. During the first half of that decade about 160,000 barrels of sperm whale oil were brought to New England's ports each year (Starbuck 1878). As the number of sperm whales dwindled and competition from coal gas and kerosene increased, whaling rapidly declined.

Matches, invented in China in the late sixth century, made it easier to kindle fires and light lamps. Previously, tinder had to be ignited. The earliest matches, slender pinewood sticks impregnated with sulfur, reached Europe only in the early sixteenth century. Modern safety matches, which incorporated red phosphorus in the striking surface, were first introduced in 1844 and soon captured most of the market (Taylor 1972).

Transportation and Construction

Thus, and not otherwise, it began with mechanics, which began first to reveal to the world how to cultivate the fields and how to subject the horse and the ox to the yoke, ... how to hitch them to wagons ... and by pulling to take us ... to the far ends of the earth, bringing back ... food, merchandise, and other great loads like rocks, wood, trees, and the materials that carpenters, marble workers and architects use in their work.

—Agostino Ramelli, *Le Diverse et Artificiose Machine* (1588)

The preindustrial evolution of traditional transportation and construction shows a highly uneven pattern of remarkable advances and prolonged stagnation or even decline. The ordinary sailing ships of the late eighteenth century were greatly superior to the best vessels of classical antiquity, both in terms of speed and in their ability to sail much closer into the wind. Similarly, well-upholstered coaches, drawn by efficiently harnessed horses and sitting on good steel springs, offered an incomparably more comfortable and faster ride than travel on horseback or in unsprung carts.

But even in the richest European countries typical roads were hardly better, and often much worse, in the eighteenth century than they had been during the last centuries of the Roman empire. Moreover, the Athenian architects who designed the Parthenon and the Roman masons who finished the Pantheon possessed skills hardly inferior to those of their successors who built late Baroque palaces and churches. Shortly afterwards, everything changed rapidly. The steam engine and cheap cast iron and steel revolutionized both transportation and construction.

Moving on Land

Now the first stage of the work was to dig ditches and to run a trench in the soil between them. Then this empty ditch was filled up with the foundation courses and a watertight layer or binder and a foundation was prepared to carry the pavement. ... Finally the pavement should be fastened by pointed blocks and held at regular distances by wedges.

—Statius describing the building of Via Domitiana (90 a.d.)

Walking and running were the two natural modes of human locomotion in all preindustrial societies. Energy costs, average speeds, and maximum daily distances have always depended primarily on individual fitness and the prevailing terrain. The energy cost of walking increases both below and above the optimum speed of 5 or 6 kilometers per hour. Uneven surfaces, mud, or deep snow raises the cost of walking by up to 25 or 35 percent. The energy cost of walking uphill is a function of both the gradient and the speed, and detailed studies show nearly linear increases across a broad range of speeds and inclines.

Running requires power outputs of between 700 and 1400 watts, equivalent to ten or twenty times the basal metabolic rate. Although this energy expenditure is relatively high, humans can virtually uncouple this cost from speed (Carrier 1984). Record performances in running have been improving steadily during the twentieth century (Ryder et al. 1976), and they are undoubtedly well above the best historic achievements. But there is no shortage of outstanding examples of long-distance running in many traditional societies. Pheidippides' fruitless run from Athens to Sparta just before the battle of Marathon in 490 B.C. was, of course, the prototype of great running endurance. He covered 240 kilometers in just two days to request Sparta's aid in the war against the Persians only to obtain a refusal.

Horseback riding offered a much more powerful and hence much faster means of personal transport. A fit animal could carry a rider up to 50 or 60 kilometers a day. A good rider who changed horses could cover more than 100 kilometers a day in emergencies. For many years scholars placed the beginnings of horseback riding in the Asian steppes during the middle of the second millennium B.C. New research suggests that it may have begun much earlier, around 4000 B.C. among the people of the Sredni Stog culture in today's Ukraine (Anthony et al. 1991). The still inconclusive evidence for this theory is based on the difference between the premolars of feral and domestic horses: The teeth of animals that were bitted show distinctive fractures and bevelling on micrographs.

Riding a horse has always presented a major physical challenge. The most efficient horseback-riding position requires the rider to put his or her center of gravity forward and low. The jockey's crouch ("monkey on a stick") is the best example of this technique. Curiously, this optimum was irrefutably established only before the end of the nineteenth century by Federico Caprilli (Thomson 1987). The forward-low position, used in the most exaggerated version in modern showjumping, differs radically from riding styles portrayed in historical sculptures and images. For a variety of reasons riders sat too far back, and their center of gravity was too high for their horses to move most efficiently (see Figure 4.15). Classical riders were even more disadvantaged because they did not have stirrups.

Nomadic archers did well without stirrups. Riding smaller horses and shooting arrows with powerful compound bows, they represented formidable and highly mobile fighting forces centuries before the introduction of stirrups (see A4.5). But without the secure support of stirrups, a fighter clad in armor could not even easily mount a large horse. And once on the horse's back with his lance or heavy sword, he could not fight effectively. Stirrups diffused westward through Eurasia after the third century. Only their universal adoption in early medieval Europe made armored riding, fighting, and jousting possible.

The simplest way of transporting loads is to carry them. Where roads were absent people could often do better than animals. Naturally, people had to carry

FIGURE 4.15 Simplified drawings of riding positions throughout history. These drawings, based on art and manuscripts from the periods represented, show: (*top, l. to r.*) an ancient Greek rider without stirrups, holding on by his knees; a mounted Asian archer as seen by a Han dynasty artist; a Mongol rider attacking Japanese troops in 1274 on Kyushu; (*bottom, l. to r.*) Albrecht Dürer's knight (from *The Knight, Death and the Devil*), engraved in 1513; an early eighteenth-century Spanish rider (from Francois de la Guerniere's *Ecole de Cavallerie*); and a nineteenth-century English rider (from Theodore Gericault's 1821 oil painting *The Derby of Epsom*). Even flying jockeys were sitting upright! *Sources:* Based on reproductions in Baskett (1980), Dent (1974), and Smith (1964). Not drawn to scale.

lighter loads than animals did, but human flexibility in loading, unloading, moving on narrow paths, and scrambling over difficult terrain often more than compensated for this weaker performance. Similarly, donkeys and mules with panniers were often preferred to horses. They were more nimble on narrow paths, and their harder hooves and lower water needs gave them greater resilience and endurance in difficult environments.

The most efficient method of carrying is to place the load's center of gravity above the carrier's own center of gravity—but balancing a load on one's head is not always practical. In many cultures, people have carried loads from poles or wooden yokes slung over their shoulders. Long-distance transfers, especially in difficult terrain, are best accomplished with backpacks fastened by good shoulder or head straps. Nepalese Sherpas are generally acknowledged as the best practitioners of this trade. In carrying the mountaineering supplies for climbing expeditions, they can move between 30 and 35 kilograms (close to half of their body weight) up to the base camp and carry nearly 20 kilograms up the steeper slopes in the rarer air above.

Roman *saccarii,* the laborers who reloaded Egyptian grain at Ostia harbor from ships to barges, carried sacks of 28 kilograms over short distances. The traditional Chinese sedan chair also required heavy lifting: Two men carrying one person shared the load prorating to no less than 25 to 40 kilograms per carrier. These loads corresponded to between 35 and 65 percent of the carrier's body weight; usual walking speeds under such loads did not exceed 5 kilometers per hour. In relative terms, people were better carriers than animals. Typical loads for animals were only about 30 percent of an animal's weight (that is, about 50 to 150 kilograms) on level ground and 25 percent of the animal's weight in the hills.

Massed applications of human labor aided by simple mechanical devices could accomplish some astonishingly demanding tasks. Undoubtedly the most taxing transport tasks in traditional societies were those requiring the delivery of large building stones or finished components to construction sites. Every old high culture quarried, moved, and emplaced large stones (Heizer 1966). A few ancient images offer first-hand illustrations of how this work was accomplished. Certainly the most impressive of these is an Egyptian painting from el-Bersheh cave dated to 1880 B.C. The scene portrays 166 men dragging a colossus on a sledge. The path is being lubricated by a worker pouring either oil or water from a vessel (see Figure 4.16). Because lubrication could cut the sledge friction by about half, their massed labor, reaching peak power inputs of more than 30 kilowatts, could move a 50-tonne load. Yet a number of preindustrial societies surpassed such efforts.

Inca builders used enormous irregular stone polygons with smoothed sides fitted to amazing precision. Pulling a 140-tonne stone, the heaviest block at Ollantaytambo in southern Peru, up a ramp required the coordinated force of about 2400 men (Protzen 1986). The peak power of this group must have been around 600 kilowatts, but scholars know nothing of the logistics of such an enterprise. How were more than 2000 men harnessed to pull in concert? How were they arranged to fit into the confines of narrow (6 to 8 meters) Inca ramps? Building sites from ancient Brittany are equally enigmatic. How did the people handle the Grand Menhir Brise (Niel 1961), at 340 tonnes the largest stone erected by a European megalithic society?

With the invention of the wheel, men could move loads far surpassing their body weight. The Chinese wheelbarrow, which centered the load right above the wheel's axle, could carry loads of more than 150 kilograms. European barrows, with their eccentric front wheel, usually carried no more than 60 to 100 kilograms.

Friction is an important factor in land transport. Road surface and vehicle design can thus impede or facilitate efficient travel. On a smooth, hard, dry road, for example, a force of only about 30 kilograms is needed to wheel a 1-tonne load. A loose, gravelly surface may easily call for five times as much draft. On sandy or muddy roads the multiple can be seven to ten times higher.

Axle lubricants—mutton or beef tallow and plant oils—have been used at least since the middle of the second millennium B.C. During the first century B.C.,

Celtic bronze bearings had inner grooves that contained cylindrical wooden rollers (Dowson 1973). Chinese rolling bearings may be of even greater antiquity. Ball bearings, however, are firmly documented for the first time only in early seventeeth-century Europe.

Roads in ancient societies were mostly just soft tracks turning seasonally into muddy ruts or dusty trails. The Romans, starting with *Via Appia* (the road from Rome to Capua) in 312 B.C., invested a great deal of labor and organization into an extensive network of hard-topped roads (Sitwell 1981). Well-built Roman *viae* were topped with gravel concrete, cobblestones, or slabs set in mortar. By Diocletian's reign (285–305), the Roman system of trunk roads, *cursus publicus*, covered some 85,000 kilometers. This enterprise consumed at least 1 billion labor days. This large total prorates to easily manageable requirements over the centuries of ongoing construction (see A4.6) In Western Europe, the Roman achievements in road building were decisively surpassed only during the nineteenth century, and in the eastern regions of the continent, only during the twentieth century.

Islam had no roads comparable to the Roman cursus, but its far-flung parts were connected by much-traveled caravan routes, which, technically, were mere tracks (Hill 1984). Pack camels replaced wheeled transport in the arid region between Morocco and Afghanistan before the Muslim conquest, largely owing to economic imperatives (Bulliet 1975). Pack camels were not only more powerful and faster than oxen but also had greater endurance and longevity. They could move over rougher ground, subsist on inferior forage, and tolerate longer spells of feed and water shortages. These economic advantages were strengthened with the introduction of the North Arabian saddle sometime between 500 and 100 B.C. The saddle provided an excellent arrangement for riding and carrying loads and allowed caravans to displace carts.

Incas, consolidating their empire between the thirteenth and fifteenth centuries, built an impressive road network through corvée labor. Its length totaled about 40,000 kilometers, including 25,000 kilometers of all-weather roads crossing culverts and bridges and equipped with distance markers. There were two main royal roads. The one winding through the Andes had a stone surface. Its width ranged from up to 6 meters on river terraces to just 1.5 meters cut through solid rock (Kendall 1973). The other, an unsurfaced coastal link, was about 5 meters wide. Neither road had to support wheeled vehicles, just caravans of people and pack llamas. The llamas carried 30 to 50 kilograms each and covered less than 20 kilometers a day.

During the Qin and Han dynasties the Chinese built an extensive road system totaling about 40,000 kilometers (Needham et al. 1971). This system was less extensive than the contemporary Roman cursus, both in total length and road density per unit area, and less sturdily built. Chinese roads, constructed by tamping rubble and gravel with metal rammers, provided a more elastic but less durable surface than the best Roman roads. Ox-drawn carts and wheelbarrows carried most of the goods, and people were moved in two-wheeled carts and sedan chairs. Although these roads supported an excellent messenger service that survived the decline of the Han, the landborne transport of goods and people generally deteriorated. In some parts of the country, however, the development of efficient canal transportation more than made up for this decline.

The first documented vehicles come from Uruk around 3200 B.C. They had heavy solid-disk wheels of up to 1 meter in diameter that were made of three or more dowelled and mortised planks held together with wooden battens. The subsequent diffusion of the wheel across different European cultures was remarkably rapid (Piggott 1983). Some early wheels rotated about a fixed axle; others turned together with the axle. Later craftsmen developed much lighter, free-turning spoked wheels (early in the second millennium B.C.) and four-wheeled vehicles with a pivoting front axle. This combination made sharp turns possible.

The speeds and capacities of animate land transport were low. Inefficiently harnessed horses traveling on poor roads were slow even when pulling relatively light loads. Maximum specifications restricted loads on Roman roads of the fourth century to 326 kilograms for horse-drawn wagons and up to 490 kilograms for the slower ox-drawn post carriages (Hyland 1990). On good roads, passenger horse carts could travel 50 to 70 kilometers per day, heavier horse-drawn wagons covered 30 to 40 kilometers per day, and oxen could travel 15 to 20 kilometers per day. Men with wheelbarrows would cover about 10 to 15 kilometers per day. Of course, much longer distances were covered by messengers changing fast horses. The recorded maxima on Roman roads go up to 380 kilometers per day.

These low speeds and low capacities translated into excessive costs. This reality is well illustrated by the figures given in Diocletian's *edictum de pretiis*. In A.D. 301 it cost more to move grain just 120 kilometers by road than to ship it from Egypt to Ostia, Rome's harbor. And after the Egyptian grain arrived at Ostia, just some 20 kilometers away from Rome, it was reloaded onto barges and moved against the Tiber's stream rather than hauled over land by ox-drawn wagons.

Similar limitations persisted in most societies well into the eighteenth century. At the beginning of the century, it was cheaper to import many goods to England by sea than to have them have carried by pack animals on poor roads from the country's interior. Travelers described English roads as "barbarous," "execrable," "abominable," and "infernal" (Savage 1959). Rain and snow made poorly laid-out soft dirt or gravel roads impassable for long stretches of time. In many cases they were so narrow that only pack traffic could travel them.

The roads were hardly better on the continent. By the middle of the eighteenth century, roads that could carry heavy horse-drawn wagons were either very poor (in France and Germany) or virtually nonexistent (in large parts of eastern Europe). Coach horses, harnessed in teams of four to six animals, lasted on the average less than three years. Fundamental improvements came only after 1750 (Ville 1990). Initially Europeans widened their roads and established better drainage; later they surfaced them with durable finishes (gravel, asphalt, and concrete).

With this development, heavy European horses could finally demonstrate their excellent hauling capabilities. By the mid-nineteenth century the maximum allowable French load was increased to nearly 1.4 tonnes, about four times the late Roman limit.

In urban transportation horses reached the peak of their importance during the railway age, between the 1820s and the end of the nineteenth century (Dent 1974). Although the railways were taking over long-distance shipments and travel, horse-drawn transport dominated in the rapidly growing cities of Europe and North America. Most railway shipments had to be collected and distributed by horse-drawn vans, wagons, and carts. These vehicles also delivered food and raw materials from the nearby countryside. Greater urban affluence brought many more private coaches and hansoms, cabs, omnibuses (first introduced in London in 1829), and various delivery wagons (see Figure 4.17).

Stabling these animals in mews and storing enough hay and straw to feed them made enormous demands on urban space. At the end of Queen Victoria's reign, London had some 300,000 horses. City planners in New York were thinking about setting aside a belt of suburban pasture to accommodate large herds of horses between the peak demands of rush-hour transport. The direct and indirect energy cost of urban horse-drawn transport—in terms of feeding, stabling, grooming, shoeing, harnessing, and driving the horses and removing their wastes to periurban market gardens—were among the largest items on the energy balances of late-nineteenth-century cities. This equine dominance ended rather abruptly. Electricity and internal combustion engines were becoming practicable just as the numbers of urban horses were rising to record totals during the 1890s. In less than a generation, horse-drawn city traffic was largely displaced by electric streetcars, automobiles, and buses.

Curiously, European and American mechanics came up with a practical version of the most efficient human-powered locomotion, the modern bicycle, at around the same time (Whitt and Wilson 1982). For generations bicycles were clumsy, even dangerous, contrivances that had no chance of mass adoption as vehicles of convenient personal transport (see Figure 4.18). Rapid improvements came only during the 1880s. John Kemp Starley and William Sutton introduced bicycles with equal-sized wheels, direct steering, and diamond-shaped frames of tubular steel. These designs have been closely followed by virtually all twentieth-century machines. The evolution of the modern bicycle was largely complete with the addition of pneumatic tires and back-pedal brakes in 1889.

Improved bicycles equipped with lights, various load carriers, and tandem seats became common for commuting, shopping, and recreation in a number of European nations, most prominently in the Netherlands and Denmark. The prevalence of the bicycle in Communist China is well known; in fact, China is still the world's largest producer of bicycles, with more than 35 million new machines built annually during the late 1980s (State Statistical Bureau 1991).

FIGURE 4.17 An engraving showing the high density of horse-drawn traffic in late nineteenth-century London. *Source:* Reproduced from *The Illustrated London News,* November 16, 1872.

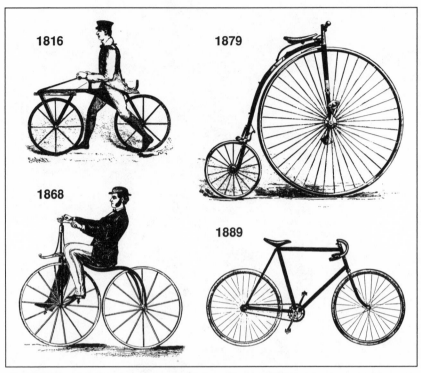

FIGURE 4.18 Early bicycles. The bicycle was a surprisingly late invention and advanced rather slowly. Riders had to push themselves on Baron von Drais's clumsy 1816 *draisine*. Pedals were first applied to the axle of the drive wheel in 1855, and this advance led to the velocipedes of the 1860s. Subsequent design regression led to huge front wheels and plenty of accidents; the first safe and efficient modern bicycle was not developed until the 1880s. *Source:* Adapted from Byrn (1900).

Shipping

They now went sayling in the OCEAN vast,
Tacking the snarling Waves with crooked Bill:
The whispring Zephyr breath'd a gentle Blast,
Which stealingly the spreading Canvas fills:
With a white foam the Seas were overcast,
The dancing Vessels cutting with their Keels
The Waters of the Consecrated DEEP,
Where Prometheus's Flocks their Rendezvous keep.

Luis de Camoes, *Os Lusiadas* (1572)

Ships propelled by human labor were common in antiquity. Many traditional oared vessels were ingeniously designed in order to integrate the efforts of tens

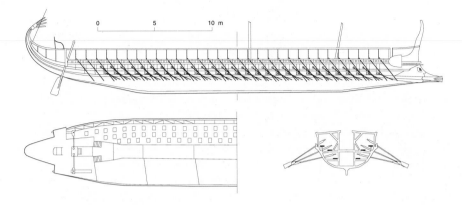

FIGURE 4.19 Complete side view, partial plan, and cross-section of a reconstructed Greek trireme, *Olympias*. Six files arranged in a V shape accommodate 170 rowers, and the topmost oars have their pivots on outriggers. *Source:* Based on Coates (1989).

and even hundreds of oarsmen. Naturally, pulling heavy oars for prolonged time periods was very strenuous; when done in confined quarters below deck it was extremely exhausting. Our admiration of the complex designs and organizational mastery of large oared ships must be tempered by the realization of the human suffering they exacted.

Ancient Greek oared ships have been particularly well studied (Morrison and Williams 1968). The vessels that took the Greek troops to Troy, the *penteconteres,* were crewed by 50 oarsmen and for brief intervals could receive useful power inputs of up to 7 kilowatts. The triple-tiered *trieres,* better known by their Latin name *triremes,* the best-performing classical warships, were powered by 170 rowers (see Figure 4.19). When crewed by skilled, healthy oarsmen they could be propelled for brief periods by more than 20 kilowatts of power, enough to produce speeds of up to 20 kilometers per hour.

Even when moving at more common top speeds of 10 to 15 kilometers per hour, the highly maneuverable triremes were powerful fighting machines. Their bronze ram could hole the hulls of enemy ships with devastating effect. In one of the decisive battles of Western history, a Greek force defeated a large Persian fleet at Salamis (480 B.C.) in just this fashion (Coates 1989). Triremes were also the most important warships of republican Rome. A working, full-scale reconstruction of the ship was finally accomplished during the 1980s (Morrison and Coates 1986).

Larger ships—*quadriremes, quinquiremes,* and so on—followed in rapid succession after Alexander's death (323 B.C.). As there is no indication that any of these ships had more than three tiers of oarsmen, two or more men presumably powered every oar. The largest known ship of antiquity was the *tessarakonter,* constructed during Ptolemaios Philopator's reign (222–204 B.C.). The 126-meter-long ship was to carry more than 4000 oarsmen and nearly 3000 troops; theoretically it could be propelled by 5 megawatts of power. But its weight, including heavy catapults, made it virtually immovable—a costly shipbuilding miscalculation.

In the Mediterranean, large oared vessels retained their importance well into the seventeenth century. At that time the largest Venetian galleys had 56 oars, each crewed by 5 men (Lane 1934). Interestingly, large Maori dugout canoes were oared by almost as many warriors (up to 200). Consequently, it is safe to conclude that the general limits of aggregate human power in sustained rowing applications were between 12 and 20 kilowatts.

Other ships powered by human labor included vessels with pedals or treadmills. During the Sung dynasty the Chinese built increasingly larger paddle-wheel warships powered by up to 200 men (Needham et al. 1965). In Europe, smaller tugs powered by 40 men turning capstans or treadmills appeared in the middle of the sixteenth century.

Animate power also constituted the principal prime mover for canal boats and barges. Canals were important catalysts of economic development in the core area of China (in the lower basin of the Huang He and on the North China Plain) during the Han dynasty as well as in subsequent eras (Chi 1936; Needham et al. 1971). By far the longest and most famous of these transport arteries was *da yunhe,* the Grand Canal. Its first section was opened in the early seventh century, and its completion in 1327 made it possible to move barges from Hangzhou to Beijing. This is a latitudinal difference of 10° and an actual distance of nearly 1800 kilometers.

Early canals used inconvenient double slipways where oxen hauled boats to higher levels. The invention of the canal pound-lock in 983 made it possible to raise boats safely and without wasting water. The progression of locks raised the Grand Canal's highest point to just over 40 meters above sea level. Chinese canal boats were pulled by gangs of laborers or by oxen or water buffaloes. In Europe, where canals reached their greatest importance during the eighteenth and nineteenth centuries, horses were the leading prime movers.

European transportation canals, an unmistakable import from China, were first constructed in North Italy during the sixteenth century. The 240-kilometer French Canal du Midi was completed by 1681. The longest continental and British links came only after 1750, and the German canal system actually postdated railroads (Ville 1990). Canal barges moved large quantities of raw materials and import commodities for expanding industries and growing cities, and they also took out their wastes. They handled a large share of European traffic just before the introduction of railways and for a few decades thereafter (Hadfield 1969).

Horses or mules on adjoining towpaths could pull a loaded barge at a speed of about 3 kilometers per hour. The mechanical advantages of barges are obvious. On a well-designed canal, a single heavy horse could pull a load of 30 to 50 tonnes—an amount an order of magnitude higher than even the strongest horse could manage on the best hard road. Steam engines gradually replaced barge-towing animals, but many horses still worked on small canals during the 1890s.

In contrast to canal barges and warships, long-distance sea-borne transportation was dominated by sailing ships from the very beginning of high civilization. The history of sailing ships may be understood primarily as a quest for higher efficiencies in harnessing the kinetic energy of the wind. Sails alone could not do this, but they were obviously the key to nautical success. They are basically fabric aerofoils designed to maximize lift force and minimize drag (see A4.7).

Square sails set at right angles across a ship's long axis provided efficient energy conversion only with the wind astern. Roman ships pushed by northwesterlies could make the Messina-Alexandria run in just six to seven days—but the return could take forty to seventy days! There were substantial seasonal differences, and all travel often ceased during the winter (for example, shipping between Spain and Italy was closed between November and April). Because there were so many variables, it is almost impossible to say what speeds were typical (Duncan-Jones 1990). Naturally, longer voyages against the wind were primarily due to lengthy course changes. The ships had to make repeated downwind turns, which took much time and labor.

There is no doubt that all of the Old World's ancient ships were rigged with square sails and that radically different designs did not emerge for many centuries. Ships with fore-and-aft rigging had sails aligned with the vessel's long axis. The masts were pivots—the sails swung around them to catch the wind—and sailors could change the ship's direction easily. The earliest fore-and-aft rigging most likely came from Southeast Asia in the form of a rectangular canted sail. Modifications of this ancient design were eventually adopted both in China and, through India, in Europe.

Characteristic batten-strengthened Chinese lug sails originated in the second century B.C., and canted square sails became common in the Indian Ocean during the third century B.C. The latter was a clear precursor of the triangular (lateen) sail that was so typical of the Arab world after the seventh century. In Europe, sailing closer to the wind only became possible in late medieval times with the combination of square rigging and triangular sails. Gradually, European ships were rigged with a larger number of loftier and more easily adjustable sails (see Figure 4.20). Sailing ships became efficient wind energy converters only with better and deeper hull designs, stern-post rudders (in China since the first century A.D., in Europe only a millennium later), and magnetic compasses (in China after 850, in Europe around 1200).

This combination was made almost irresistibly powerful by the addition of accurate, heavy guns. The gunned ship, developed in Western Europe during the fourteenth and fifteenth centuries, launched an era of unprecedented long-distance expansion. In Carlo Cipolla's (1966) apt characterization, the ship "was essentially a compact device that allowed a relatively small crew to master unparal-

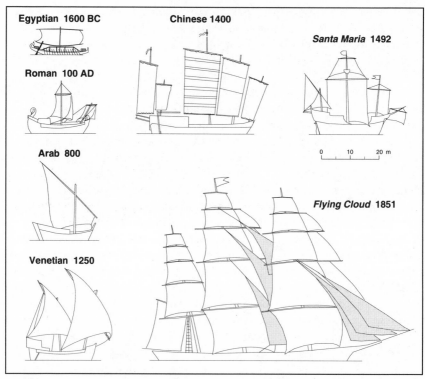

FIGURE 4.20 Evolution of sailing ships. Ancient and classical Mediterranean cultures used square-rigged ships (*top left*). Triangular sails (*bottom left*), dominant for centuries in the Indian Ocean and throughout the Arab world, were later adopted by Europeans. A large seagoing junk from Jiangsu (*top center*) typifies efficient Chinese designs with batten sails. Columbus's *Santa Maria* (*top right*) had square sails, a foretopsail, a lateen on mizzen, and the spritsail under the bowsprit. *Flying Cloud* (*bottom right*), a famous mid-nineteenth-century record-breaking U.S. clipper, was rigged with triangular jibs fore, spanker aft, and lofty main royal and skysails. *Sources:* Simplified outlines drawn to scale from historic images in Armstrong (1967 and 1969), Daumas (1969), and Needham et al. (1971).

leled masses of inanimate energy for movement and destruction. The secret of the sudden and rapid European ascendancy was all there" (p. 137).

Starting in the early fifteenth century the simplest ships of this innovative design carried the audacious Portuguese sailors on long voyages. First they sailed along the western coast of Africa (reaching the mouth of the Senegal in 1444, the equator in 1472, and Angola in 1486); in 1497 Vasco da Gama rounded the Cape of Good Hope and crossed the Indian Ocean to India. Before the century's end three Spanish ships had crossed the Atlantic (1492), and in 1519 Magellan traversed the Pacific and his *Victoria* completed the first circumnavigation of the world. European colonization of distant territories and a huge increase in maritime trade soon followed.

Sailing ships improved dramatically between medieval times and the industrial age, both in terms of speed and in terms of their load capacities. Rich historical records enable us to chart the progress. As with waterwheels, late medieval performances showed no substantial gains over antique ratings. Although the Romans built ships with capacities of more than 1000 tonnes, their standard cargo vessels carried less than 100 tonnes. More than a millennium later, Europeans embarked on their explorations with ships nearly as small. In 1492 Columbus's *Santa Maria* carried 165 tonnes, and Magellan's ship rated a mere 85 tonnes. A century later (1599), the vessels of the Spanish Great Armada averaged 515 tonnes. By 1800 British ships in the Indian fleet had capacities of about 1200 tonnes.

Roman cargo ships could not go faster than 2 or 2.5 meters per second; the best mid-nineteenth-century clippers, however, could surpass 9 meters per second. In 1853, the *Lightning*, a Boston-built ship with a British crew, logged the longest daily run under the sail: It traveled 803 kilometers at an average speed of 9.3 meters per second (Wood 1922). And in 1890, *Cutty Sark*, perhaps the most famous tea clipper, ran 6000 kilometers in thirteen consecutive days and averaged 5.3 meters a second (Armstrong 1969). There is no reliable way in which to calculate the total energy needed to move individual ships on long voyages; nor is it possible to know how much energy was harnessed annually by a nation's merchant or military fleets. According to Unger (1984), sailing ships harnessed as much energy as windmills did during the Dutch Golden Age. Nevertheless, this amount was equivalent to less than 5 percent of the country's huge peat consumption (see A4.8).

Building

> *One part is city, one part orchard, and one part claypits. Three parts including the claypits make up Uruk.*
>
> —Description of Uruk in the Sumerian epic *Gilgamesh*
> (beginning of the second millennium b.c.)

Behind the enormous variety of building styles and ornaments are only four basic structures: walls, columns, beams, and arches. Human labor, aided by a few simple tools, could create all of these from the three basic building materials of the preindustrial world—timber, stone, and brick. Preindustrial peoples cut down trees and roughly shaped them with axes and adzes. They quarried stone with hammers and wedges and shaped it with chisels, and they made bricks from readily available alluvial clays. Some constraints did exist, however. Shortages of large trees limited the use of timber in many regions, and the expense of transporting stone restricted the choices largely to local varieties.

Sun-dried mud bricks, common throughout the Middle East and Mediterranean Europe, were the least energy-intensive building blocks. Mixtures made from loams or clays, water, chaff or chopped straw, and sometimes dung or sand, were compacted, rapidly shaped in wooden molds (up to 250 pieces per hour), and left to dry in the sun. Dimensions ranged from the chunky, square Babylonian pieces ($40 \times 40 \times 10$ centimeters) to the slimmer, oblong ($45 \times 30 \times 3.75$ centimeters) Roman bricks. Because they were poor heat conductors, mud bricks provided good passive air conditioning in hot arid climates. They had also an important mechanical advantage: Building a mud-brick vault requires no wood beams for support (Van Beek 1987). With suitable clays and enough available labor, the bricks could be produced in prodigious quantities.

Burnt bricks were used in ancient Mesopotamia, and later they were common in both the Roman empire and Han China. For centuries most of the bricks were fired in unenclosed piles or pits, but this method wasted fuel and resulted in uneven baking. Later brick makers used regular mounds or stacks. These could reach temperatures up to 800 °C and yielded a more uniform product with much higher efficiency. Completely enclosed horizontal kilns assured better consistency and higher combustion efficiency. They had properly spaced flues to channel the smoke and domed roofs to reflect the rising hot gases. In sixteenth-century Europe, when bricks started to replace wattle-and-daub or timber studding and became more common in foundations and walls, the need for wood and charcoal to supply the kilns increased proportionately.

Regardless of the principal materials used, preindustrial structures demonstrate a skillful integration of labor. Large numbers of men and animals accomplished tasks that appear extraordinarily demanding even by modern standards. Moreover, many of these structures were built very quickly. Khufu's pyramid, the largest stone structure ever built, took only twenty years to build; the Parthenon, perhaps the world's most ideally proportioned building, took only fifteen years (447–432 B.C.); and the magnificent Hagia Sophia in Constantinople, a boldly vaulted Byzantine church turned into a mosque, was completed in less than five years (527–532)!

Several types of large construction projects stand out. By far the best known are various ceremonial structures, above all funerary monuments and places of worship. The most remarkable structures in the first group—pyramids and tombs—are distinguished by their massiveness, while temples and cathedrals combine monumentality with complexity and beauty. These tombs and temples express the universal human striving for permanence, perfection, and transcendence. Other projects were more utilitarian. Among these, aqueducts are noteworthy because of their length and the combination of canals, tunnels, bridges, and inverted siphons.

It is impossible to prepare energy accounts for ancient construction projects—even the energy costs of medieval projects are not easy to estimate. Nevertheless, it is clear that such projects required huge and sustained energy flows, long-range planning, outstanding organization, and large-scale labor mobilization. The construction of the Egyptian pyramids, the grandest structures of the ancient world, had to be a marvel of organization. Building them required planning, efficient

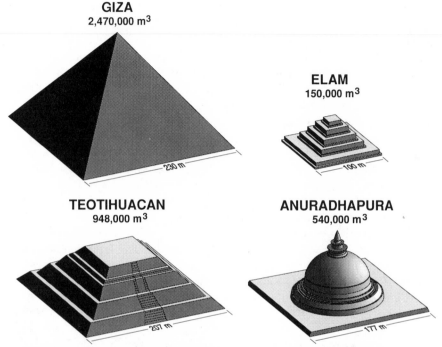

GIZA
2,470,000 m³

ELAM
150,000 m³

TEOTIHUACAN
948,000 m³

ANURADHAPURA
540,000 m³

FIGURE 4.21 Scaled oblique views, dimensions, and volumes of Khufu's pyramid at Giza, the Choga Zanbil *ziggurat* at Elam, the Sun Pyramid at Teotihuacan, and the Jetavana *stupa* at Anuradhapura. Sources: Based on Bagenal and Meades (1980), Bandaranayake (1974), Porada (1965), and Tompkins (1976).

grand-scale logistics, effective supervision and servicing, and admirable technical skills. And yet, the technical skills used to build these monuments are almost completely obscured and the energy requirements can only be roughly estimated.

The largest pyramid, the pharaonic tomb of Khufu of the fourth dynasty, embodies the best all of these qualities (see Figure 4.21). Built of nearly 2.5 million stones weighing on the average about 2.5 tonnes each, this mass of over 6 million tonnes within a volume of 2.5 million cubic meters was assembled with remarkable precision and speed (Barber 1900). The Egyptian hieroglyphic and pictorial record, so rich in many other aspects, provides no contemporary or later descriptions or images informing us about its construction. Standard Egyptological explanations assume that laborers quarried the core stones at the site and brought the facing stones across the Nile (Lepre 1990). A unique image from Deir el-Bahari attests to the fact that very large stones were transported on boats. Other sources reveal that two Karnak obelisks, each 30.7 meters long, were carried on a barge (63 meters long) pulled by about 900 oarsmen in thirty boats (Naville 1908).

Herodotus described a grand inclined plane of polished stone that extended from the Nile riverbank to the Giza plateau; according to his account, gangs of workers pulled sleds loaded with stone blocks over this structure. He also explained that the blocks were lifted from ground level by "contrivances made of short timbers," but most Egyptologists have believed that clay, brick, and stone ramps were used in erecting the pyramid itself (Grimal 1992). This explanation is highly unlikely (Hodges 1989). A single inclined plane would have to be completely rebuilt after finishing every layer of stonework. Moreover, the volume of stone that went into this plane would have far surpassed that needed for the pyramid itself. Ramps encircling the pyramid would have been narrow and difficult to negotiate, but this problem could have been eased by pivoting ropes used to pull stones at right angles around corner posts.

One pyramid-building experiment used clay ramps with embedded wooden planks. Using the technique described above, workers could easily move wooden sleds loaded with large stones up progressively steeper slopes (Hadingham 1992). Proponents of lifting have suggested that the work could have been done with the help of levers or simple but ingenious machines. Peter Hodges (1989) argues for the simplest method—using wooden levers to lift the stone blocks and then rollers to emplace them. Depending on various assumptions, the construction may have required anywhere from several thousand to nearly 100,000 laborers (see A4.9).

Mesoamerican pyramids, especially those at Teotihuacan (built during the second century A.D.) and Cholula, are also quite imposing. The tallest of these, Teotihuacan's flat-topped Pyramid of the Sun, was probably just over 70 meters including the temple. It was much easier to construct than the three stone structures at Giza. The pyramid's core is made up of earth, rubble, and adobe bricks; only the exterior was faced with cut stone. The stone was anchored by projecting catches and plastered with lime mortar (Baldwin 1977). Still, its completion could have required up to 10,000 laborers for more than twenty years. Other massive, solid, pyramid-like structures that required large labor forces include the Mesopotamian stepped temple towers (*ziggurats*) built after 2200 B.C. and Buddhist monuments housing relics (*stupas*) in India, Ceylon, and Southeast Asia.

There is less mystery about such classical Mediterranean temples as the Parthenon. Its architraves weigh almost 10 tonnes and had to be lifted 10.5 meters. Men could perform that task readily, albeit slowly, by turning capstans or windlasses or treading drums to power a crane. Very similar cranes were used nearly two millennia later for building cathedrals, the most elaborate structures of the European Middle Ages (see Figure 4.22). Large animal teams transported building materials to a site, but construction relied solely on human labor. Craftsmen used saws, axes, hammers, chisels, planes, augers, compound pulleys, and treadwheels for lifting timber, stones, and glass (Fitchen 1961; Wilson 1990).

Although some cathedrals were completed quickly (Chartres took only twenty-seven years; the original Notre Dame de Paris took thirty-seven years), construction was frequently interrupted by money shortages, wars, and disputes. Building

FIGURE 4.22 Examples of the principal structural elements of medieval cathedrals: high vaults (*left;* here the nave of the St. Etienne's in Caen), large ornamental stained-glass windows (*top right;* from Reims), and flying buttresses (*bottom right;* from the choir of Notre Dame in Paris). *Source:* Adapted from Wilson (1990).

a cathedral often stretched over many decades and even centuries. Much of the labor was seasonal, but the construction typically engaged hundreds of full-time workers—lumbermen, quarrymen, wagon drivers, carpenters, stonemasons, and glassworkers—for one or two decades. The total energy investment was thus two orders of magnitude smaller than in pyramid building, with peak labor flows of only a few hundred kilowatts.

It is well documented that extensive waterworks, including dams, canals, and bridges, existed in Jerusalem, Mesopotamia, and Greece. But Roman achievements are certainly the best-known examples of bold engineering solutions to urban water supply problems. Virtually every sizable Roman town had a well-planned water supply. This accomplishment was surpassed only by European water projects during industrialization (Smith 1978). The city of Rome's aqueducts were especially impressive (see Figure 4.23). Pliny, in *Historia Naturalis,* called them the most remarkable achievement anywhere in the world.

Starting with Aqua Appia in 312 B.C., the water supply system eventually comprised eleven lines that extended for a total length of almost 500 kilometers (Ashby 1935; Hodge 1992). During the second century, when this system satisfied the needs of more than 1 million people, it probably delivered no less than 1 million cubic meters (1 billion liters) per day. A daily per capita average of close to 1 cubic meter of water was an enormous flow even when compared to modern supply rates. Italian domestic water use in the late 1980s was less than half a cubic meter a day (World Resources Institute 1992). Equally impressive was the scale of Rome's underground sewage canal system, which included *cloaca maxima* arches of about 5 meters in diameter.

Throughout the Roman empire aqueducts included a number of common structural elements. Starting from springs, lakes, or artificial impoundments, water channels had rectangular cross-sections and were built of stone slabs or concrete lined with fine cement. Channels with the usual gradient of no less than 1 in 200 followed slopes in order to avoid tunneling whenever possible. Where an underground course was unavoidable, the channel could be accessed from above by shafts. Only in valleys too long to skirt or too deep for simple embankments did the Romans resort to bridges. No more than about 65 kilometers of Rome's aqueducts were carried on arches, which they sometimes shared. The Augustan bridges at Gard (more than 50 meters high), Merida, and Taragona are the finest examples of this art. Workers were continuously cleaning and repairing these channels, tunnels, and bridges, which were often threatened by erosion.

When crossing a valley would have required a bridge taller than 50 or 60 meters, Roman engineers opted for an inverted siphon. Its pipes connected a header tank on one side of the valley with a slightly lower-lying receiving tank on the opposite side (Hodge 1985). A bridge crossed the stream at the valley bottom. These structures—especially the high-pressure pipes, which could withstand up to 18 atmospheres—required large amounts of lead. The metal had to be transported over often considerable distances—and at high energy cost. For example, the nine siphons in the Lyon water supply used about 15,000 tonnes of the metal.

Metallurgy

The first is the furnace or other agent which contains the fire and the metal brought together. The second is the wood or charcoal which is the necessary and proper nutriment of the fire, from whose virtues are derived the greater or lesser forces which, according to their kind, are ready to effect what is required.

—Vannoccio Biringuccio, *De la Pirotechnia* (1540)

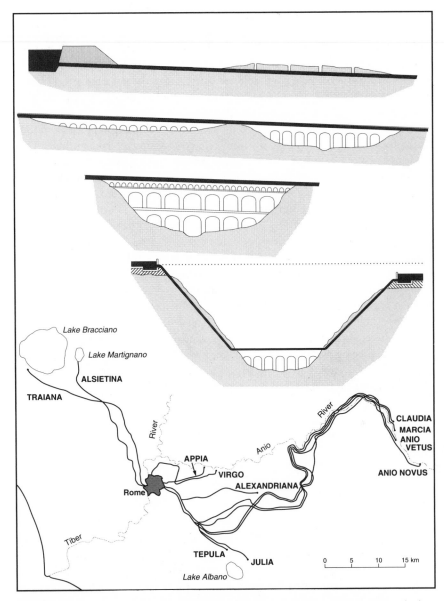

FIGURE 4.23 Roman aqueducts carried water from rivers, springs, lakes, or reservoirs by combining at least two or three of the following structures (*top to bottom*): shallow rectangular channels running on a foundation; tunnels accessible by shafts; embankments pierced by arches; single- or double-tiered arched bridges; and inverted lead-pipe siphons to take water across deep valleys. Rome's aqueducts, supplying about 1 million cubic meters of water a day, formed an impressive system built over a period of more than 500 years. *Sources:* Based on Ashby (1935) and Smith (1978). Aqueduct slope is exaggerated.

The beginnings of all old high cultures are marked by the use of color (nonferrous) metals. Besides copper, early metallurgists also recognized tin (combined with copper in bronze), iron, lead, mercury, and the two precious elements, silver and gold. Mercury is, of course, a liquid at ambient temperatures. Gold's relative scarcity and softness made it suitable for coins and ornaments but precluded its use for mass-produced items. Although much more abundant than gold, silver was also too rare for mass-produced items. Lead and tin are soft; when unalloyed with other metals they were usually only useful for such things as pipes and food containers. Only copper and iron were relatively abundant and durable. They possessed, especially when alloyed, great tensile strength and hardness. These properties made them the only two practical choices for mass-produced items. Copper and bronze dominated the first two millennia of recorded history, while iron and its alloys are the dominant metals even today.

Charcoal fueled the smelting of both nonferrous and iron ores as well as the subsequent refining and finishing of crude metals and metallic objects. Laborers spent long hours mining and crushing the ore, cutting trees and making charcoal, building furnaces, and charging, casting, refining, and forging the metals. In many societies—ranging from sub-Saharan Africa to Japan—metallurgy remained solely manual labor until the introduction of modern industrial methods. In Europe, and later in North America, animals and, above all, water power took over such repetitive, exhausting tasks as crushing ore, pumping water from mines, and forging metal. The availability of wood, and later also the accessibility and reliability of water power to energize larger bellows and hammers, were thus the key determinants of metallurgical progress.

Nonferrous Metals

Find the copper tablet-box, slip loose the ring-bolt made of bronze.

—Narrator in the Sumerian epic *Gilgamesh*
(beginning of the second millennium b.c.)

Copper tools and weapons bridged the stone and iron eras. The first uses of copper, datable to the sixth or fifth millennium B.C., did not involve any smelting. Ancient people used simple tools to shape pieces of naturally pure metal or used an annealing process (alternate heating and hammering). Smelting and casting became common after the middle of the fourth millennium B.C. in a number of regions with rich and relatively accessible oxide and carbonate ores (Forbes 1972). Numerous copper objects—rings, chisels, axes, knives, and spears—were left behind by the early Mesopotamian societies (before 4000 B.C.), predynastic Egypt (before 3200 B.C.), the Mohenjodaro culture in the Indus valley (2500 B.C.), and the ancient Chinese (after 1500 B.C.).

The copper-mining centers of antiquity included Egypt's Sinai peninsula,

North Africa, Cyprus, the regions covered by today's Syria, Iran, and Afghanistan, the Caucasus, and Central Asia. Later, Italy, Portugal, and Spain produced copper. Given the metal's relatively high melting point (1083 °C), production of pure copper was rather energy-intensive. Early cultures used wood or charcoal to reduce the ore in clay-lined pits; later ones used simple, low shaft clay furnaces with a natural draft. The first clear evidence of bellows comes from Egypt and dates from the sixteenth century B.C., but they almost certainly go back further. Workers refined impure metals by heating them in small crucibles then cast them into stone, clay, or sand molds. They fashioned castings into utilitarian or ornamental products by hammering, grinding, piercing, and polishing them.

Much higher technical skills were necessary for producing the metal from abundant sulphide ores. Workers first crushed and roasted the ores in heaps or furnaces in order to remove the sulfur and other impurities (antimony, arsenic, iron, lead, tin, and zinc) that affected the metal's properties. For millennia, workers used hammers to crush ores by hand, a practice common in Asia and Africa until the twentieth century. In Europe, waterwheels and horses harnessed to whims gradually took over this work. Roasting the crushed ores required relatively little fuel. After smelting the roasted ore in shaft furnaces, workers smelted and resmelted the coarse metal (only 65 to 75 percent copper) to produce nearly pure (95 to 97 percent) blister copper. This product could be further refined by oxidation, slagging, and volatilization. The fuel requirements of the entire sequence were high.

Using Robert Forbes's (1972) energy analysis of Roman copper production, it is possible to calculate the extent of deforestation caused by the Rio Tinto furnaces (see A4.10). As with so many other techniques, Roman achievements remained unsurpassed for another 1500 years. Copper smelting in late medieval times hardly differed from the Rio Tinto practices (Biringuccio 1540; Agricola 1556).

Bronze was the first practical alloy and was chosen by Christian Thomsen for his now classical division of human evolution into the Stone, Bronze, and Iron ages (Thomsen 1836). This division is highly generalized. Some societies (most notably Egypt before 2000 B.C.) went through a pure copper era, whereas others (above all those in sub-Saharan Africa) moved directly from stone to iron. The first bronzes were accidents resulting from the inadvertent smelting of copper ores containing tin. Later they were produced by cosmelting of copper and tin ores, and only after 1500 B.C. were they made by smelting the two metals together. With its low melting point of a mere 231.97 °C, tin was produced with relatively little charcoal from its crushed oxide ores. The total energy cost of bronze was thus lower than that of pure copper, but the alloy possessed superior properties.

As the shares of tin in bronze varied anywhere from 5 to 30 percent (and, consequently, the melting point of bronze ranged between 750 and 900 °C), it is impossible to speak of a typical bronze. An alloy preferred for casting guns, composed of 90 percent copper and 10 percent tin, was superior to the best cold-drawn copper: Its tensile strength and hardness was about 2.7 times higher (Lyman 1961).

Bronze thus made it possible to produce the first good metallic axes, chisels, knives, and bearings as well as the first reliable swords. Bronze bells usually were 25 percent tin.

Brass, the other historically important copper alloy, combined the element with zinc. In these alloys, copper made up anywhere from 50 to 85 percent of the total. Like bronze production, brass production uses less energy than the smelting of pure copper (zinc's melting point is only 419 °C). The higher the zinc content, the higher the alloy's tensile strength and hardness. For a typical brass, these factors are about 1.7 times higher than for cold-drawn copper. The higher zinc percentages do not reduce the alloy's malleability or its corrosion resistance. The first uses of brass date to the first century B.C. The alloy was widely used in Europe only during the eleventh century and became common only after 1500.

An unusual copper alloy provided the metal of everyday use for the pre-Incan Mochica, Sican, and Chimus societies of South America between the first and fifteenth centuries (Shimada and Merkel 1991). These societies used inefficient methods to smelt this copper-arsenic alloy in simple, shallow furnaces. The labor requirements were high. Because the charcoal charge was inadequate and the blast insufficient for producing liquid metal, the workers had to recover small droplets of metal by grinding the slag.

Iron and Steel

A certain quantity of iron ore is given to the master, out of which he may smelt either much or little iron. He being about to expend his skill and labour on this matter, first throws charcoal into the crucible, and sprinkles over it an iron shovelful of crushed iron ore mixed with unslaked lime. Then he repeatedly throws on charcoal and sprinkles it with ore, and continues this until he has slowly built up a heap; it melts when the charcoal has been kindled and the fire violently stimulated by the blast of the bellows.

—Georgius Agricola, *De re metallica* (1556)

Iron gradually replaced copper and bronze. The Mesopotamians produced small iron objects during the first half of the third millennium B.C., but ornaments and ceremonial weapons became more common only after 1900 B.C. No society used iron extensively until after 1400 B.C., and the metal became truly abundant only after 1000 B.C. Egypt's iron era dates from the seventh century B.C., China's from the sixth. African ironmaking is also ancient. The metal was never smelted by any New World society. Iron smelting was necessarily bound up with large-scale production of charcoal. Iron melts at 1535 °C. An unaided charcoal fire can reach 900 °C, but a forced air supply can raise its temperature to nearly 2000 °C. Charcoal thus fueled all iron ore smelting in every traditional society except for China, which also used coal starting in the Han dynasty. Although charcoal remained the

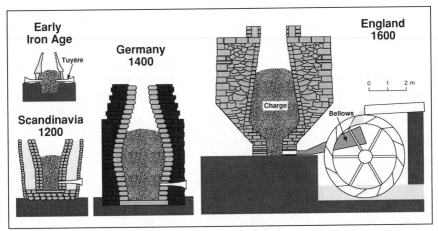

FIGURE 4.24 Evolution of ironmaking furnaces. These scaled drawings show the histori-cal development of furnaces, from ancient pit hearths (which still produced iron in Africa during this century) to the mid-sixteenth-century English predecessor of tall charcoal-fueled blast furnaces. *Sources:* Based on Haaland and Shinnie (1985), Forbes (1956), and Crossley (1981).

primary energy source for three millennia, new techniques increased the effi-ciency of iron production substantially.

The earliest iron producers smelted the crushed ore in shallow pits, often lined with clay or stone (see Figure 4.24). These primitive hearths were commonly lo-cated on hilltops to maximize natural draft. Later producers used a few narrow clay tubes (tuyeres) to deliver a blast from a small leather bellows into the hearth. They erected simple clay walls, from just a few decimeters to more than a meter high, to contain the smelting. Archaeologists have unearthed thousands of these temporary structures throughout the Old World, from India to the Iberian pen-insula and from northern Europe to Central Africa (Maddin 1988; Haaland and Shinnie 1985).

These simple hearths did not produce liquid iron, only a spongy mass (usually between 25 and 70 kilograms) of small iron globules embedded in slag and cin-ders, known as "bloom." Only by repeatedly reheating and hammering the bloom could an iron worker produce a lump of tough and malleable wrought iron. This metal, which was less than 0.1 percent carbon and had a melting point of 1535 °C, was used in making objects and tools ranging from nails to axes. Liquid iron was first produced by Chinese craftsmen during the Han dynasty (207 B.C.–A.D. 220). The Chinese furnaces, built with refractory clays and often strengthened by vine cables or heavy timbers, eventually reached just over 5 meters. They could take a charge of nearly 1 tonne of iron ore and produced cast iron in two tappings a day. The Chinese used ores with a high phosphorus content, which lowered the melt-ing point of the metal, and invented double-acting bellows that delivered a strong air blast. These advances supplied critical ingredients for China's early success (Needham 1964).

Later Chinese craftsmen packed coal around the batteries of tube-like crucibles containing the ore. They also used a larger bellows powered by waterwheels. Using interchangeable molds, they were mass producing iron tools, thin-walled cooking pots and pans, and statues before the end of Han dynasty (Hua 1983). As with so many early Chinese inventions, however, there were few substantial improvements in the centuries that followed, and China's small blast furnaces did not start the lineage of today's huge structures.

These structures slowly evolved from European shaft furnaces. From a simple Catalan forge to the rock-lined *osmund* furnaces of Scandinavia and the *Stückofen* of Styria, these furnaces became more and more efficient. Higher stacks and better shapes lowered fuel consumption. Higher temperatures and longer contact between the ore and the fuel made liquid iron production possible. European blast furnaces originated most likely in the lower Rhine valley region just before 1400. Blast furnaces produce cast iron, also called pig iron, an alloy that is 1.5 to 5 percent carbon and that cannot be directly forged or rolled. Its tensile strength is no higher than that of copper (and it can be up to 55 percent weaker), but it is two to three times harder than copper (Lyman 1961).

The number of blast furnaces in Europe grew steadily during the sixteenth and seventeenth centuries. The most notable improvement of that time was a larger bellows made of wood and bull hides. After 1620 came double bellows, operated alternatively by the cams on the waterwheel axle, as well as a gradual elongation of the stack. Both of these trends soon ran into limits imposed by the maximum power of waterwheels and the physical properties of charcoal. By 1750 the largest waterwheels could deliver up to 7 kilowatts of useful power. But during the summer months there was often not enough water to generate the maximum output. The main disadvantage of charcoal is its high friability: It easily crushes into dust under heavy loads; thus, blast-furnace stacks had to be less than 8 meters tall (Sexton 1897). Before 1800 both of these limitations were removed—the first by Watt's steam engine, the second by the use of coke.

According to Otto Johannsen (1953), medieval bloomery hearths used 3.6 to 8.8 kilograms of fuel per kilogram of ore. Even with good ores that had a metal content of about 60 percent, bloomery smelting required at least 8 kilograms and as much as 20 kilograms of charcoal per kilogram of hot metal. By the end of the eighteenth century, typical charcoal to metal ratios were around 8:1, and by 1900 this rate was down to just around 1.2:1. The best rates were as low as 0.77:1 in Swedish charcoal furnaces (Campbell 1907; Greenwood 1907). A good late nineteenth-century charcoal-fueled furnace thus required only about one-tenth the energy that its medieval counterpart consumed!

The high energy requirements of pre-1800 charcoal-fueled smelting inevitably caused extensive deforestation around furnace sites. A typical early eighteenth-century English furnace required about 1600 hectares of trees for a sustainable supply (see A4.11). In the early 1700s, English iron production consumed about 1100 square kilometers of forest annually; a century later, U.S. iron production consumed more than twice that amount (see A4.12).

Not surprisingly, communities surrounded by traditional iron mills and forges rapidly found themselves in a desperate situation. In 1548 the anguished inhabitants of Sussex were already wondering what number of towns would be likely to decay if the iron mills and furnaces were allowed to continue. They feared they would have no wood to build houses, watermills, wheels, barrels, piers, and hundreds of other necessities—and they asked the king to close down many of the mills (Straker 1969). The limiting role of energy in traditional iron smelting was thus unmistakable. When a single furnace could strip each year a circle of forest with a radius of about 4 kilometers, it was easy to appreciate the cumulative impact of scores of furnaces over a period of many decades.

This effect was necessarily concentrated in wooded, mountainous regions. There, close to their fuel supply, the producers could keep the radius of animal-drawn charcoal transportation to a minimum and power furnace and forge bellows by waterwheels. Proximity to the ore was also important, but because the ore charge was only a fraction of the weight of the charcoal, it was easier to transport. Deforestation was the inevitable environmental price paid for nails, axes, horseshoes, mail shirts, lances, guns, and cannonballs. The early expansion of iron making led to an energy crisis in Britain during the seventeenth century as the supply of domestic wood dwindled. That situation was further aggravated by the country's burgeoning shipbuilding industry, which also had high timber demands (Thomas 1986).

Although iron was relatively abundant in many preindustrial societies, steel was available only for special uses. Like cast iron, steel is also an alloy, but it is only 0.15 to 1.5 percent carbon and contains small amounts of other metals (mainly nickel, manganese, and chromium). Steel is superior to cast iron and copper alloys: Depending on its composition, its tensile strength is 2 to 4 times higher than copper's and its hardness is between 3 and 5.3 times higher (Lyman 1961).

Some simple ancient smelting techniques could produce relatively high-quality steel, but only in small amounts. For example, traditional East African steelmakers used low, cone-shaped furnaces (less than 2 meters high) built of slag and mud over a pit of charred grass. Using a method apparently known from the first centuries A.D., eight men operated a goatskin bellows connected to ceramic tuyeres. With this procedure, the charcoal-fueled furnace reached temperatures above 1800 °C (Schmidt and Avery 1978).

Most preindustrial societies followed one of two effective routes toward steel: They either carburized wrought iron or decarburized cast iron. The first, and older, technique consisted of prolonged heating of the metal in charcoal, which resulted in the gradual inward diffusion of carbon. Without further forging this method produced a hard steel layer over a core of softer iron. This was a perfect material for plowshares—or for body armor. Repeated forging distributed the absorbed carbon fairly evenly and produced excellent sword blades. Decarburization, the removal of carbon from cast iron by oxygenation, was practiced in China during the Han dynasty. The process produced metal for such exacting applications as chains for suspension bridges.

The spreading availability of iron and steel led gradually to a number of profound social changes. Iron saws, axes, hammers, and nails speeded up house construction and improved its quality. Iron kitchenware and a variety of other utensils and objects, ranging from rings to rakes and from grates to graters, made it easier to cook and to run a household. Iron horseshoes and plowshares were instrumental in advancing the intensification of cropping. On the destructive side, warfare was revolutionized—first by flexible chain-mail suits, helmets, and heavy swords, later by guns, iron cannonballs, and increasingly more reliable firearms. These trends were greatly accelerated by the introduction of coke-based iron smelting and the emergence of the steam engine.

APPENDIXES

A4.1 Work, Force, and Distance

Work is done when a force—provided by either animate or inanimate prime movers—changes a body's state of motion. The magnitude of work is equal to the product of exerted force and the distance covered in the direction in which the force acts. Naturally, the same amount of work can be accomplished by applying a greater force over a shorter distance or a smaller force over a longer distance.

A4.2 Power of Overshot Waterwheels

The potential energy of water (in joules) is equal to the product of its mass (in kilograms), head (in meters) and the acceleration of gravity (9.8 meters per second squared [m/s^2]). Consequently, an overshot wheel bucket containing 0.2 m^3 of water (200 kg) poised 3 m above the discharge channel has a potential energy of roughly 6 kJ. With a water flow rate of 400 kg/s the wheel would have a theoretical power of nearly 12 kilowatts (kW). The useful mechanical power of such a machine would have ranged from less than 4 kW for a heavy wooden wheel to well over 9 kW for a carefully crafted and well-lubricated nineteenth-century metal machine.

A4.3 Wind Energy and Power

Average wind speed is proportional to the height above the ground raised to the power of 0.14. This means, for example, that the speed of wind 20 meters above the ground will be about 22 percent higher than that of wind at a height of 5 meters. The kinetic energy of 1 cubic meter (m^3) of air (in joules) is equal to $0.5DV^2$, where D is air density (about 1.2 kg near the ground), and V is average wind speed (in meters per second). The power of the wind (in watts) is the product of wind energy, wind speed (V), and the area perpendicular to the wind direction swept by the machine's blades (A, in m^2), or 0.5DAV^3. As the wind power goes up with the cube of the average speed, doubling the speed increases the available power eightfold (2^3). Thus, locations with sustained brisk winds are superior to those with mild winds.

Preindustrial societies could capture only the wind flowing within 30 to 35 meters above the ground, and most of the windmills built over centuries had spans of less than 10 meters. Wind flows also exhibit large temporal and spatial variations. Even in windy places, annual wind speed averages will fluctuate by up to 30 percent. Because of terrain irregularities, shifting a machine's location by just 30 to 50 meters may easily increase or decrease the average speed by one-half. Modern machines generating electricity can be flexibly located at such exceptionally windy sites as California passes or the Danish west coast, but traditional locations of human settlements favored sheltered places with frequent calms. Given the limited capacities of preindustrial land transport, it was impractical to locate windmills at the best available sites; as a result, the mills often had to stay motionless.

Naturally, no wind machine can extract all of the available wind power: This would require the machine to completely stop the airstream so that the air could accumulate at the point of conversion! The maximum extractable power is equal to 16/27, or nearly 60 percent, of the kinetic energy flux (Krenz 1976). Actual performance is close to 50 percent for modern wind machines, and it averaged between 20 and 30 percent for preindustrial windmills. Consequently, an eighteenth-century wood-and-canvas tower mill with a blade diameter of 20 m could theoretically extract about 112 kW when turning in wind with a velocity of 10 m/s—but it actually delivered less than 40 kW of useful power.

A4.4 Energy Content of Biomass Fuels

Biomass Fuel	Water Content (percent)	Energy Density of Dry Matter (MJ/kg)
Hardwoods	15–50	16–19
Softwoods	15–50	21–23
Charcoal	<1	28–30
Crop residues	5–60	15–19
Dry straws	7–15	17–18
Dried dung	10–20	8–14

Source: Derived from Smil (1983).

A4.5 Kinetic Energy of Arrows

A typical lightweight arrow weighing just 20 grams and launched by a good archer from a compound bow could travel at a speed of up to 40 meters per second (m/s) (Pope 1923). The kinetic energy of this projectile would be 16 joules. This figure appears to be very low only until one remembers that the impact is basically punctiform. Consequently, flint- or metal-tipped arrows could easily go through a coat of mail when shot from distances of up to 40 to 50 meters, and they could kill unprotected men from a distance of more than 200 meters.

A4.6 Energy Cost of Roman Roads

If the average Roman road was just 5 meters wide and 1 meter deep, workers would have to remove some 800 million cubic meters (m^3) of earth and rock and emplace about 425 million m^3 of sand, gravel, concrete, and stone to construct the roadbeds, embankments, and ditches for 85,000 kilometers of trunk roads. If each worker handled only 1 m^3 of building materials per day, the tasks of quarrying, cutting, crushing, and moving stones, excavating sand for foundations, ditches, and roadbeds, preparing concrete and mortar, and laying the road would have added up to about 1.2 billion labor days.

If maintenance and repair eventually tripled this requirement, the grand total over 600 years of construction would have averaged out to 6 million labor days per year, an equivalent of some 20,000 full-time construction workers. At, say, 800 kilojoules per hour, this amount of labor would represent a net annual energy investment of nearly 5 terajoules (TJ) and a peak power of more than 4 megawatts (MW).

A4.7 Sails and Sailing Near the Wind

When wind strikes a sail, the difference in pressure generates two forces: lift and drag. The direction of the lift is perpendicular to the sail, and drag acts along the sail. With wind astern, or nearly so, lift force will be much stronger than drag force and a ship will make good progress. With wind on the beam, or slightly ahead, the force pushing the vessel sideways is stronger than the force propelling it forward. If the ship tries to steer even closer to the wind, the drag will surpass the lift and the vessel will be pushed backwards.

Maximum capabilities for sailing near the wind have advanced by more than 100° since the earliest days of sailing. Early Egyptian square-sailed ships could manage only a 150° angle (wind astern, only 30° off the desired course), whereas medieval square rigs could proceed slowly with the wind on their beam (90°), and their post-Renaissance successors could move at an angle of just about 80° into the wind. Sailing closer to the wind only became possible with the use of asymmetrical sails mounted more in line with the ship's long axis and capable of swiveling around their masts. Ships combining square sails with triangular mizzens could manage 62°, and fore-and-aft rigs (including triangular, lug, sprit, and gaff sails) could come as close as 45° to the wind. Modern yachts come very close to 30°, the aerodynamic maximum.

The only way to circumvent the earlier limits was to proceed under the best manageable angle and keep changing the course. Square-rigged ships had to resort to wearing, making a complete downwind turn. Ships with fore-and-aft sails were tacking, turning their bows into the wind and eventually catching it on the opposite side of the sail.

A4.8 The Contribution of Sailing Ships to Dutch Energy Use

The amount of energy needed to move individual ships on long voyages and the aggregate annual contributions of wind power harnessed by merchant or military fleets are impossible to calculate with information on tonnages and speeds alone. Critical variables—hull designs, sail areas and cuts, cargo weights, and utilization rates—are far too heterogeneous to allow for estimation of meaningful averages.

Still, using a set of assumptions, Richard Unger (1984) attempted to calculate the contribution of sailing ships to energy use during the Dutch Golden Age. He ended up with an annual total of roughly 6.2 megawatts of power during the seventeenth century. For comparison, this is almost exactly equal to the total energy output from all Dutch windmills as estimated by DeZeeuw (1978)—and only a small fraction (less than 5 percent) of the country's huge peat consumption.

But such quantitative comparisons mean little: No amount of peat would have made the trips to the East Indies possible. The useful energy gained from peat was almost certainly less than half of its gross heat value, and, of course, peat is a limited and rapidly depletable resource, whereas wind is an abundant and renewable one. Comparisons of aggregate power outputs thus do not make any more sense than comparisons of individual conversion efficiencies (sailing ships versus peat stoves, for example).

A4.9 Energy Cost of Pyramid Construction

Average labor rates can be used to derive plausible estimates of the labor needed for quarrying and lifting the stones to construct the Egyptian pyramids. With an average daily rate of just 0.25 cubic meters (m^3) of stone per man, only about 1500 quarrymen—using copper chisels and dolerite mallets and working 300 days a year—would be needed to cut 2.5 million m^3 of stone in twenty years. Even if three times as many masons were needed to square and dress the stones, the total labor force in quarries would be just around 5000 men.

Peter Hodges (1989), allowing three years for the dressing of casing stones in place, calculated that 125 teams working for seventeen years could have jacked all the stones into position. If he is correct, only about 1000 permanent workers were needed for lifting stones. But there is no clear way to estimate the labor requirements for transporting stones to the site.

In his *Histories*, Herodotus reported that 100,000 laborers were needed for three months a year for twenty years. Kurt Mendelssohn (1974), using basic physical requirements, estimated the total at 70,000 seasonal laborers and perhaps as many as 10,000 permanent masons. These commitments translate to anywhere from 100 million to 200 million labor days. With ten-hour days and average net energy inputs of 800 kilojoules per hour (kJ/h), this labor would represent a total human energy investment of between 0.8 and 1.6 petajoules (PJ) and a peak power of between 12 and 15 megawatts (MW).

By comparison, the construction of the Anu *ziggurat* at Warqa required at least 1500 men working ten hours a day for five years (Falkenstein 1939), while the largest Anuradhapura *stupa*, built of some 200 million mostly rough-laid bricks, needed only about 600 corvée laborers working 100 days a year for fifty years (Leach 1959).

A4.10 Fuelwood Needs for Roman Copper Smelting at Rio Tinto

The average fuel need in copper smelting was about 90 kilograms of wood per kilogram of metal, and about 60,000 tonnes of copper were recovered by the Romans. Smelting this much copper would have required wood from about 40,000 hectares of natural deciduous forest. If the area surrounding the Rio Tinto furnaces had been completely wooded, this demand would represent every tree within a radius of 11 kilometers. The Rio Tinto slag heaps amassed some 900,000 tonnes of slag during the era of Roman mining; centuries of copper smelting on Cyprus (starting around 2600 B.C.) left behind more than 4 million tonnes of slag. Clearly, ancient smelting was a major cause of deforestation in the Mediterranean region, as well as in Transcaucasia and Afghanistan, and local wood shortages limited the extent of smelting.

A4.11 Fuel Needs of an Eighteenth-Century English Blast Furnace

Early eighteenth-century English blast furnaces worked only from October to May and during that time their average output was just 300 tonnes of pig iron (Hyde 1977). If the furnaces used just 8 kilograms (kg) of charcoal per kg of iron and 5 kg of wood per kg of charcoal, a single furnace would use 12,000 tonnes of wood annually. After 1700 nearly all accessible natural forest growth was gone and the wood was cut in ten- to twenty-year rotations from coppicing hardwoods. The annual harvestable increment would be between 5 and 10 tonnes per hectare (t/ha). A medium productivity of 7.5 t/ha would have required about 1600 ha of coppicing hardwoods for perpetual operation. For comparison, a large but inefficient seventeenth-century English furnace in the Forest of Dean required about 5300 ha of coppice growth, whereas the smaller ironworks at Wealden required around 2000 ha for each furnace-forge combination (Crossley 1990).

A4.12 Energy Needs in British and American Iron Production

In 1720, 60 British furnaces produced about 17,000 tonnes of pig iron. Producing this amount of pig iron required 40 kilograms (kg) of wood per kg of metal, or about 680,000 tonnes of trees. Forging 12,000 tonnes of bars required another 2.5 kg of charcoal per kg of bars, or 150,000 tonnes of trees, for a total annual consumption of some 830,000 tonnes of charcoaling wood. With an average annual wood productivity of 7.5 t/ha the British iron industry would have required about 1100 square kilometers (km^2) of forests and coppiced growth in 1720.

The earliest available pig iron total for the United States is for 1810, when producing about 49,000 tonnes of the metal used about 2 million tonnes of wood, or about 2600 km^2 of forest each year, an area of roughly 50 × 50 km. If the country's iron making had continued to run on charcoal a century later, even the much lower charging rates (1.2 kg of charcoal, or just 5 kg of wood, per kg of metal) would have put an enormous strain on U.S. forests. The annual metal output of about 25 million tonnes would have required about 170,000 km^2 of forestland by 1910. This area is equal to a square with each side equivalent to the distance between Philadelphia and Boston!

5

Fossil-Fueled Civilization

I see two paramount themes in the history of the past century: the growth of human control over inanimate forms of energy; and an increasing readiness to tinker with social institutions and customs in the hope of attaining desired goals.
—William H. McNeill, *The Rise of the West* (1963)

FUNDAMENTALLY, no civilization on Earth can be anything but a solar society depending on the flow of the sun's radiation to maintain a habitable biosphere and to energize photosynthesis, the material foundation of all higher life. All preindustrial societies used solar energy both directly and indirectly. The sun's direct radiation provided light and heat, and its conversions included food, feed, and wind and water flows. These flows are almost immediate transformations of insolation, whereas food and feed incorporate solar radiation with delays ranging between a few months and a few years. And it also takes only a few years for animate prime movers to reach their working ages: Children in traditional societies started to work as soon as they were five or six years old, draft animals even sooner. Only when mature trees were burned or turned into charcoal was the use of solar radiation postponed for decades.

Fossil fuels are also transformations of solar radiation: They arose from a slow alteration of biomass by pressure and heat. The ages of these fuels range from at least a few thousand years for young peat to hundreds of millions of years for hard coals. Whereas preindustrial societies tapped virtually instantaneous energy flows, converting only a tiny fraction of a practically inexhaustible energy income, modern civilization depends on extracting prodigious energy stores and is depleting finite fossil energy that could not be replenished even in 100 million years.

By turning to these rich stores we have created societies that consume unprecedented amounts of energy. This transformation brought enormous advances in agriculture and fast growth to industrializing economies. Abundant food and industrial mass production spread a measure of affluence even to rapidly growing populations and led to the emergence of a global market, growing personal mobility, and an age of greater accessibility to ever richer stores of information. By far the most troubling negative consequences of these developments are the deep socioeconomic disparities between high and low users of energy, the geopolitical tensions based on the uneven distribution of energy resources, and the development of weapons of mass destruction.

Even if the modern civilization is able to avoid a large-scale thermonuclear conflict, it faces profound uncertainties. The most worrisome challenge is the widespread environmental degradation. This decline stems from the extraction and conversion of both fossil fuels and nonfossil energies, industrial production, and rapid urbanization. The cumulative effects of these changes can go beyond local and regional problems to cause destabilizing global biospheric change.

All of this is just an interlude. Unlike its predecessors, fossil-fueled civilization cannot last thousands of years. Stores of fossil energies are finite, and even their most efficient use cannot prolong their exploitation beyond half a millennium or so. Indeed, the end of fossil-fueled societies will come well before the actual physical exhaustion of coals and hydrocarbons. The rising costs and growing environmental burdens of fossil fuel use will force our descendants to turn to solar energy flows or develop new sources of energy.

The Great Transition

Nature, in providing us with combustibles on all sides, has given us the power to produce, at all times and in all places, heat and the impelling power which is the result of it.

—Sadi Carnot, *Reflexions sur la puissance ... (1824)*

In some countries fossil fuels were used, albeit in relatively small quantities, for centuries before the beginning of a rapid displacement of biomass fuels and animate labor. Coal and natural gas in China, and coal in England, are the best-known examples. Coal was generally the dominant fossil fuel during the European transition. The most notable exception concerned one of the continent's most influential economies: During the seventeenth century the Dutch Golden Age was energized largely by domestic peat (DeZeeuw 1978).

The Russian, North American, and Japanese transitions also started with coal, but unlike Europe those countries switched sooner and faster to oil and natural gas. A large number of Asian, African, and Latin American countries (even those with coal deposits) skipped the coal stage and became rapidly dependent on domestic or, more commonly, imported crude oil. And whereas most of Europe reduced its dependence on biomass fuels to very low levels during the nineteenth century, that transition is still under way in many poor countries.

If patterns differ, so do the fuels. Coals, crude oils, and natural gases have a wide range of properties (see A5.1). Through combustion, they can be used directly for cooking food, warming rooms, or smelting metals and indirectly for energizing various prime movers. The steam engine became the leading prime mover of the nineteenth century. Internal combustion engines and steam turbines started to make commercial inroads during the 1890s. Before 1950, gasoline and diesel engines became the dominant prime movers in transportation, whereas steam turbines made possible the large-scale generation of electricity.

Beginnings and Diffusion of Coal Extraction

Until the sixteenth century coal was hardly ever burned, either in the family hearth or the kitchen, at distances of more than a mile or two from the outcrops, and, even within the area thus circumscribed it was used only by the poor who could not afford to buy wood.

—J. U. Nef, *The Rise of the British Coal Industry* (1932)

The beginnings of coal utilization go back to antiquity when the Han dynasty Chinese used it in iron production (Needham 1964). European records show the important dates: the first extraction in Belgium in 1113, the first shipments to London in 1228, the first exports from the Tynemouth region to France by 1325 (Nef 1932). England was the first country to accomplish the shift from plant fuels to coal; the transition occurred during the sixteenth and seventeenth centuries after serious regional wood shortages led to rising costs for fuelwood, charcoal, and lumber (Nef 1932; Harris 1974). These shortages worsened during the seventeenth century, and they were only temporarily alleviated by higher imports of bar iron and timber (Thomas 1986). Domestic coal extraction was the obvious solution: Almost all of the country's coalfields were opened up between 1540 and 1640.

By 1650 the annual coal output had exceeded 2 million tonnes; the annual output reached 3 million tonnes in the early eighteenth century and more than 10 million tonnes by the century's end. The rising use of coal presented many technical and organizational problems. Once they had depleted the outcropping seams, mine operators developed deeper pits. Although the pits were rarely more than 50 meters deep during the late seventeenth century, the deepest shafts surpassed 100 meters shortly after 1700, 200 meters by 1765, and 300 meters after 1830. By that time daily production was between 20 and 40 tonnes per mine, compared to just a few tonnes per mine a century earlier. With the deeper pits, mine operators encountered more water, which had to be pumped. They also had to ventilate the deeper mines and hoist the coal from deeper shafts. And, of course, they had to distribute the coal to their customers. Waterwheels, windmills, and horses powered these needs. Coal mining itself was energized solely by heavy human labor.

Sometimes crawling through narrow tunnels, hewers wielded picks, wedges, and mallets to extract coal from seams. Putters filled woven baskets with coal and dragged them on wooden sledges to the bottom of the pit, where onsetters hung them on ropes. Windsmen hauled the baskets up and banksmen dumped the coal onto piles. Adult men did most of the extraction, but boys as young as six to eight years old were employed for lighter tasks. In many pits the heaviest work was done by women or teenaged girls. They had to carry coal to the surface by ascending steep ladders with heavy baskets, fastened with straps to their foreheads, on their backs. One eighteenth-century observer noted that the women were "weeping most bitterly from the excessive severity of their labour" (Ashton and Sykes 1929).

Later, after the deeper pits had been developed, horses turned the whims (much like the one pictured in Figure 1.2) for hoisting coal or pumping water. After 1650, horses and donkeys also performed some tasks underground. Horse-drawn wagons, sometimes on rails, transported the coal over short distances to rivers or harbors, where it was loaded onto canal boats or ships. By the beginning of the seventeenth century, coal was used in heating forges, firing bricks, tiles, and earthenware, making starch and soap, and extracting salt. But it could not yet be used in such endeavors as making glass, drying malt, or, most important, smelting iron because its combustion transferred impurities to the final products.

The glass-making problem was solved first, in around 1610, with the introduction of reverberating (heat-reflecting) furnaces, which heated the raw materials in closed vessels. It became possible to use coal in iron and malt production only with the availability of coke. This fuel is almost pure carbon. To prepare it, coke producers heat suitable bituminous coals in the absence of air to drive off virtually all the volatile matter. Coke was used first in small amounts in malt drying after 1640. Abraham Darby succeeded in producing pig iron with coke in 1709. Coking offered a virtually unlimited supply of a superior metallurgical fuel, but the process was initially wasteful and costly. After 1750 coke became widely used in English blast furnaces (Harris 1988). Its ability to support heavier charges led to taller stacks and larger internal volumes and outputs. In turn, these changes increased the demand for coking coal. The higher temperature of coke-smelted molten iron also brought about improvements in metal casting.

Another important indirect use of coal came with production of coal gas, or town gas (Elton 1958). The first practical installations of coal gas fixtures appeared in English cotton mills in 1805–1806. A company was chartered in 1812 to provide a centralized gas supply for London. Technical advances insured the rapid diffusion of gas lighting. Inventors developed better retorts, more efficient burners, and new techniques for removing sulfur from the gas and making small-diameter wrought iron pipes. Gas lighting did not disappear with the introduction of light bulbs. An incandescent gas-mantle patented in 1885 by Carl Auer von Welsbach enabled the gas industry to compete with electric lights for a few more decades.

During the nineteenth century coal's most important role was providing the fuel for steam engines, which were employed in countless industrial tasks as well as in land and water transport. Coal mining and steam engines reinforced each other. The fact that miners had to pump more water from deeper mines provided an impetus for developing steam engines in the first place. The increased coal production in turn led to the availability of cheaper fuel, which brought proliferation of steam engines and thus a further expansion of mining. Soon the engines also powered winding and ventilating machinery (see Figure 5.1).

Outside England, coal mining diffused rather slowly during the eighteenth century. Major outputs came first from northern France, from the Liege and Ruhr regions, and from parts of Bohemia and Silesia. North American coal extraction started to matter nationwide only during the early nineteenth century. The early dominance of British coal mining declined slowly. In 1800 the country produced more than four-fifths of the coal extracted worldwide. Its share was still over 50 percent in 1870, and it maintained its primacy until the century's end. Afterwards English production was rapidly surpassed by U.S. extraction.

Steam Engines

> *I have found Common steam engines are Imperfect 1st because the Cylinder*
> *having been Cooled by the Injection in the preceding stroke Condenses a*
> *Considerable Quantity of Steam besides what is necessary to fill it. 2dly because*
> *the Vacuum is Imperfect without the Cylinder. ... 3dly they have not hitherto been*
> *applyed to any Circular Motion.*
>
> —James Watt, an early draft of steam-engine specifications (prior to 1769)

The steam engine was the first new prime mover since the adoption of windmills. It was the first practical, economic, and reliable machine that could convert coal's chemical energy into mechanical energy. The invention was of profound importance for worldwide industrialization (Dickinson 1939; Jones 1973; von Tunzelmann 1978). Its evolution started with Denis Papin's experiments with a tiny model in 1690. Soon after Papin's toy-like machine came Thomas Savery's small steam-driven pump (about 750 watts), which operated without a piston. By 1712 Thomas Newcomen had built a 3.75-kilowatt engine to power mine pumps. Because this engine condensed steam on the underside of the piston, it had a very low efficiency (0.7 percent at best).

Subsequent improvements (mainly through the work of John Smeaton) roughly doubled this poor performance, and after 1750 Newcomen's engines were slowly spreading to English mines. In his greatest innovation, James Watt increased the engine's efficiency by adding a separate condenser (see Figure 5.2). This innovation was patented on January 5, 1769, and the patent was extended for twenty-five years by the Steam Engine Act of 1775. Watt also added an insulated steam jacket around the cylinder and an air pump to maintain the vacuum in the steam condenser. Later he introduced a double-acting engine (which allowed the piston to drive the engine on the downstroke as well as the upstroke) and a centrifugal governor to maintain a constant speed with varying loads.

Watt's superior engine and Matthew Boulton's financing made for a very successful combination. By 1800 the partnership had built about 500 engines for both reciprocal and rotary applications. The average capacity of these machines, about

FIGURE 5.1 A typical English coal mine of the steam engine era (the C Pit of the Hebburn Colliery). The engine, housed in the building with a stack, powered the winding and ventilation machinery. *Source:* Reproduced from Hair (1844).

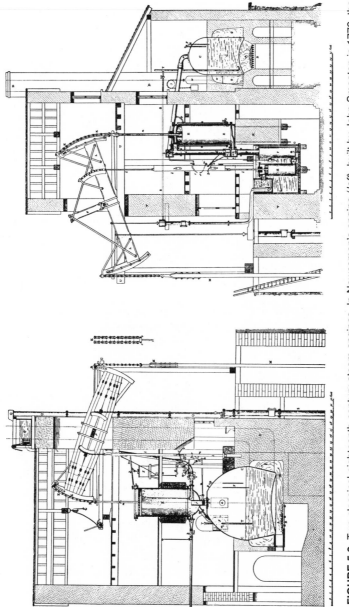

FIGURE 5.2 Two classical eighteenth-century steam engines. In Newcomen's engine (*left*), built by John Smeaton in 1772, the boiler was placed underneath the cylinder and the steam was condensed inside the cylinder. Water was injected from the pipe leading to the lower right. In Watt's engine (*right*), built in 1788, the boiler was placed in a separate enclosure, the cylinder was enveloped by an insulated steam jacket, and a separate condenser was connected to an air pump that maintained a vacuum.

Source: Reproduced from Farey (1827).

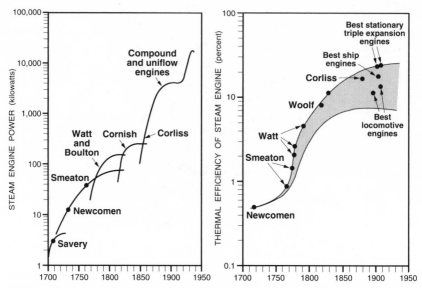

FIGURE 5.3 The rising power and improving efficiency of the best steam engines, 1700–1930. *Sources:* Plotted from data in Dickinson (1939) and von Tunzelmann (1978).

20 kilowatts, was more than five times higher than the mean for typical contemporary watermills and nearly three times higher than that for windmills. Watt's largest units (just over 100 kilowatts) matched the most powerful existing waterwheels. But the location of waterwheels was inflexible; steam engines could be sited with incomparably greater freedom.

Although Watt's inventions opened up the way for the engine's industrial success, the long term of his patent actually impeded further innovation. For reasons of safety, Watt was against using high-pressure steam. He made no attempt to develop steam-driven transportation, and he actually discouraged William Murdock, the principal installer of his engines, from doing so (Robinson and Musson 1969). After Watt's patent expired in 1800, an intense period of innovation made steam engines both more efficient and more versatile. Richard Trevithick introduced high-pressure boilers in England in 1804, and Oliver Evans did so in the United States in 1805. Other milestones included Jacob Perkins's uniflow design in 1827, George Henry Corliss's regulating valve gear in 1849, and French improvements in compound locomotive engines after the mid-1870s.

The largest machines designed during the last decade of the nineteenth century were about thirty times more powerful than those designed in 1800, and the efficiency of the best units was ten times better (see Figure 5.3). This huge performance gain, which translated into large fuel savings and less air pollution, came mainly from a hundredfold rise in operating pressures. These advances—combined with the portability of the new engines, their adaptability to many manu-

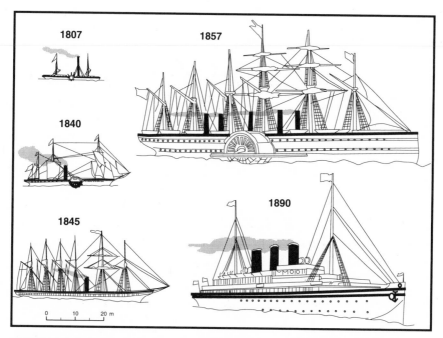

FIGURE 5.4 Evolution of the steamship. In less than a century steamships progressed from Fulton's 40-meter-long *Clermont* in 1807 (*top left*), a boat with small paddlewheels, to sleek, large, screw-propelled passenger liners. The *Normannia* (*bottom right*), launched in 1890 and serving the Hamburg–New York line, was 156 meters long and could reach nearly 40 km/hour. Subsequent designs had better engines but remained outwardly very similar. I. K. Brunel's 208-meter-long *Great Eastern* (*top right*), launched in 1857, was an exceptional giant converted from passenger service to cable-laying. *Source:* Scaled drawings based on Croil (1898).

facturing, construction, and transportation uses, and their durability—turned the machine into *the* prime mover of nineteenth-century industrialization.

A score of basic engine types gave rise to a large variety of specialized designs (Watkins 1967). The steam engine's uses soon extended to a large variety of stationary and mobile applications. It powered belt drives in countless factories and revolutionized nineteenth-century land and water transportation. The development of steamships and steam locomotives proceeded concurrently. The first steamboats were built during the 1780s in France, the United States, and Scotland, but the first commercially successful ships came only in 1802 and 1807 (Patrick Miller's *Charlotte Dundas* in England and Robert Fulton's *Clermont* in the United States, respectively; see Figure 5.4).

All early river-going ships were propelled by paddlewheels (astern or amidship), as were still fully rigged ships for sea travel. The first Atlantic crossing was the Quebec-London trip by the *Royal William* in 1833 (Fry 1896). The first west-

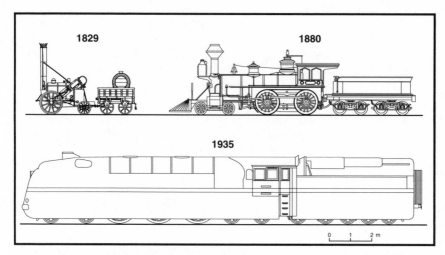

FIGURE 5.5 Notable machines of the steam locomotive age. Richard Trevithick's 1804 locomotive was the first steam-powered machine on rails. George Stephenson's 1829 *Rocket* (*top left*) introduced two features that were included in every subsequent machine: separate cylinders on each side that moved the wheels by means of short connecting rods and an efficient multitube boiler. The standard American type (*top right*) dominated U.S. railways since the mid-1850s. The streamlined German *Borsig* (*bottom*) reached 200.4 km/h in 1936. *Sources:* Based on Byrn (1900) and Ellis (1977).

ward run was the race between the paddlewheelers *Sirius* and *Great Western* in 1838, the year in which John Ericsson deployed the first successful screw propeller. Gradually, larger and faster steamships displaced sailing ships from the busiest passenger and cargo runs across the North Atlantic, and later from long-distance routes to Asia and Australia. They transported most of the 60 million emigrants who left Europe between 1815 and 1930 for overseas destinations, above all for North America (Baines 1991).

Overland steam transport also had a slow start followed by speedy diffusion. Richard Trevithick's 1804 experiment with a machine on cast iron rails inspired a number of small, private railroads. The first public railway, from Liverpool to Manchester, opened only in 1830, its train pulled by George Stephenson's *Rocket* (see Figure 5.5). A profusion of new designs brought more efficient, faster machines. By 1900 the best locomotive engines were operating at pressures up to five times higher than those possible in the 1830s and had efficiencies of over 12 percent (Dalby 1920). Speeds of more than 100 kilometers per hour became common, and during the 1930s streamlined locomotives approached, and even surpassed, 200 kilometers per hour.

But in spite of impressive improvements in the steam engine and its domination of industrial and transportation markets, the machine could not remain the leading prime mover of the twentieth century. Even the largest steam engines were not powerful enough to turn the large generators necessary in rapidly growing electric stations. The smaller engines, especially in locomotives, were still not very efficient. And all steam engines were inherently massive, and hence unsuitable for fast road transport. By 1900 the lightest steam engines weighed only about one-fifth as much per watt as in 1800—but this was still more than twice as much as the new internal combustion engines. Very soon these new engines became still lighter. But they would not have diffused so quickly without the new fuels. These liquids, refined from crude oil, had a higher energy density than coal, were cleaner to burn, and were easier to move and store.

Oil and Internal Combustion Engines

In the coal oil and camphene industries, crude distillation processes to produce oil for lighting had been developed. In the salt well industry, adequate methods of drilling had been developed. A seemingly unlimited market existed. ... The only thing lacking was, in fact, petroleum—at least petroleum in sufficient quantities.

—Sam H. Schurr and Bruce C. Netschert,
Energy in the American Economy, 1850–1975 (1960)

Large-scale crude oil extraction began in the late nineteenth century. Hydrocarbons had been well known for millennia from seepages, bitumen pools, and the "burning pillars" especially common in the Middle East. But in antiquity they were used almost solely as building materials or protective coatings. Uses requiring combustion, including the heating of Constantinople's *thermae* during the late Roman empire, were rare (Forbes 1964). In a remarkable exception, however, the Chinese burned natural gas to evaporate brines in the landlocked Sichuan province (Adshead 1992).

This procedure, which began in the Han dynasty (around 200 B.C.), was made possible by the Chinese invention of percussion drilling (Needham 1964). In this process, two to six men jumped on a lever at rhythmic intervals to raise a heavy iron bit attached to long bamboo cables from a bamboo derrick. The deepest recorded boreholes extended only 10 meters during the Han dynasty, reached 150 meters by the tenth century, and culminated in the 1-kilometer deep Xinhai well in 1835 (Vogel 1993). Natural gas was distributed by bamboo pipelines to evaporate brines in huge cast-iron pans. Some of it was also used for lighting and cooking. This Chinese practice remained isolated, and the beginning of the worldwide hydrocarbon age had to wait for two millennia.

Abraham Gesner first distilled kerosene in 1853 in London, and the fuel soon replaced expensive whale oil in lamps. This development provided the initial stimulus for the oil industry. American enterprise was its driving force. On August 27, 1859, at Oil Creek, Pennsylvania, Colonel E. L. Drake's workers penetrated 10 meters of rock and completed an oil-producing well 21 meters deep (Brantly 1971). They used a procedure similar to the ancient Chinese percussion technique, but their drill was raised and lowered by a small steam engine. Major oil discoveries followed in quick succession. Before 1900, forests of derricks stood above the oilfields around Ploesti in Romania, in Baku on the Caspian Sea, in California, in Texas (after 1887), and in Sumatra (after 1893). Before World War I, oil production had also started in Mexico (1901), Iran (1908), Trinidad (1913), and Venezuela (1914). This rapid expansion led to overproduction and declining prices.

Most oil deposits are associated with natural gas, but little of the latter fuel was used during the early decades of the hydrocarbon industry. Without compressors and steel pipes, gas could not be moved over long distances. In contrast, refined oil products are easier to handle than coal and hence cheaper to transport and to store. Moreover, their energy density is about 50 percent higher than that of standard coal. This combination makes liquid hydrocarbons excellent fuels for transportation. The invention and rapid adoption of internal combustion engines opened up a huge new market for such fuels, and the growing demand for them has governed many technical and social developments of the twentieth century.

The development of the internal combustion engine, a new prime mover burning fuel within the cylinder, proceeded very rapidly. The designs perfected just before 1900 during the first generation of its commercial use remained fundamentally unchanged for most of the twentieth century. In the mid-1800s, European inventors had conducted serious experiments and developed some prototype designs. Nikolaus Otto built the first four-stroke cycle engine, which ran on coal gas, in 1876. Commercial success came only with the invention of gasoline and diesel engines. The first light, high-speed, gasoline-powered, single-cylinder vertical engine was patented by Gottlieb Daimler in 1885. In the same year Karl Benz built the world's first car, which was powered by a much slower horizontal gasoline engine.

Daimler's engine, Benz's electrical ignition, and Wilhelm Maybach's float-feed carburetor launched the automobile industry, which is still expanding today. A light internal combustion engine also powered the first flight. The Wright brothers built a four-cylinder machine with an aluminum body and a steel crankshaft that delivered up to twice as much power as originally intended—and building that engine was only a small part of their invention (Gunston 1986).

Their achievement was so remarkable because they had to carry out the whole process of aerodynamic design and testing virtually from the beginning. Their predecessors, despite their enthusiasm, had left little to build upon in terms of practical advances. The brothers solved two crucial problems, balance and control and proper wing design, in their 1902 glider. They installed the new engine and accomplished the first flights in a heavier-than-air machine on December 17, 1903 (Wright 1953; Jakab 1990). The photograph of the first take-off above the dunes of North Carolina—with Orville piloting and Wilbur watching tensely—is certainly one of the most memorable images of the twentieth century.

At the same time that spark engines were gaining commercial success, Rudolf Diesel was developing an entirely different mode of fuel ignition. In his engine, patented in 1892, the compression ratios were between 14 and 24 (compared to just 7 to 10 for Otto engines). The fuel injected into the cylinder ignited spontaneously in the high temperatures that resulted from these compression ratios. This design does have some disadvantages. The engine is heavier and runs at a lower speed. But diesel engines are inherently more efficient than Otto engines: The best diesels can surpass efficiency ratings of 40 percent, compared to 25 percent for the best Otto engines. They can also use cheaper yet more energy-dense, heavier liquid fuels. Not surprisingly, their first commercial success came in marine propulsion where their weight made little difference.

The success of internal combustion did not end the steam era. Yet another steam-driven prime mover emerged just before the end of the nineteenth century, and its subsequent development determined the course of many twentieth-century industrial advances: steam turbine rotating generators that produced electricity in increasingly larger central stations.

Electricity

All the similar ends of the compound hollow helix were bound together by copper wire, forming two general terminations, and these were connected with the galvanometer. The soft iron cylinder was removed, and a cylindrical magnet ... used instead. One end of this magnet was introduced into the axis of the helix, and then, the galvanometer-needle being stationary, the magnet was suddenly thrust in; immediately the needle was deflected.

—Michael Faraday, *On the Induction of Electric Currents* (1831)

Electricity generation, transmission, and use represented unparalleled achievements in energy innovation. Previously, all new prime movers had been designed to be used with many existing productive arrangements. In contrast, large-scale generation of electricity required the development of a new prime mover solely dedicated to the task, and the practical use of electric current necessitated the invention and installation of a whole new system that would provide for the reliable and safe distribution and effective conversion of electricity into light, heat, and motion. A number of basic discoveries had to precede the first attempts at commercial use.

None was more important than Michael Faraday's demonstration of electromagnetic induction on October 17, 1831. Faraday showed that mechanical energy can be converted into electricity and vice versa. All electricity generators and motors are based on this principle. Z. T. Gramme took a critical step toward translating this potential into reality in 1871 when he built the first ring-wound armature dynamo. Then, in 1880, Elihu Tomson patented three-phase armature winding. But more than any other nineteenth-century innovation, this new industry was driven by one man's vision.

Thomas Alva Edison was an exceptional holistic conceptualizer whose greatest genius was to direct large-scale problem solving (Hughes 1983). This work required accurate identification of technical challenges, tenacious interdisciplinary research and development, and rapid introduction of resulting innovations into commercial use (Jehl 1937; Josephson 1959). There were other contemporary inventors of light bulbs and large generators, but only Edison had the vision, determination, and organizational talent to translate his bold ideas into practical realities. He and his coworkers developed new products in astonishingly short periods of time.

The filament of a carbonized cotton sewing thread gave off steady light in Edison's first durable, high-vacuum light bulb on October 21, 1879. The first electricity-generating plant, built by Edison's London company at Holborn Viaduct, began transmitting power on January 12, 1882. New York's Pearl Street station, commissioned on September 4 of the same year, became the first American power plant. A month after its opening it energized some 1300 light bulbs in the city's financial district.

A year later more than 11,000 lights were wired. Although the first light bulbs were very inefficient by today's standards, their performance was superior to that of any other light source then available. They were about ten times brighter than gas mantles and a hundred times brighter than candles. These huge advances in lighting were important both for industrial modernization and for quality of life. A succession of timely inventions, and the victory of alternating current, sped up subsequent electrification (Electricity Council 1973). Edison was on the losing side of the emotional "battle of the systems." He favored direct current, opposing Nikola Tesla, George Westinghouse, and Sebastian Ferranti, whose designs prevailed by the 1890s. But Edison's innovative acumen led him to cut his losses and abandon the manufacture of direct current systems (David 1991).

That decade also saw the emergence of steam turbines as prime movers in electricity-generating plants. The first power stations used large steam engines. These bulky, heavy, inefficient prime movers were abandoned soon after Charles Parsons patented a more efficient, smaller, lighter steam turbine in 1884. Parsons's company installed a 75-kilowatt turbine in 1888 and progressed to a 1-megawatt unit by 1900.

Two other inventions made notable contributions to electrification—transformers and electric motors. The first device, introduced by William Stanley in 1885, made it possible to transmit high-voltage alternating current from power plants with relatively low losses and distribute it at low voltages to households and industries (Coltman 1988). Three years later, Nikola Tesla demonstrated the first practical induction motor operating on alternating current (Cheney 1981). During the remarkable decade of the 1880s, Edison, Parsons, Stanley, Tesla, Westinghouse, and their collaborators put in place the foundations of a new industry. The subsequent expansion of this industry became one of the critical defining forces of the twentieth century. By 1900, electricity was making fast inroads not only in households and industries—including the burgeoning telecommunications industry—but also on railroads.

Technical Innovation

Power, speed, motion, standardization, mass production, quantification,
regimentation, precision, uniformity, astronomical regularity, control, above all
control—these became the passwords of modern society in the new Western style.
—Lewis Mumford, *The Myth of the Machine* (1966)

In spite of their enormous differences, fuel extraction, transportation, and electricity conversion have shared a common trend. Following their initial commercialization during the nineteenth century, each field saw a period of rapid advance. This progression was interrupted by World War I, as well as by the economic crisis of the 1930s. World War II speeded up the development of nuclear energy, jet planes, and rocket propulsion. The renewed growth of all energy industries after 1945 clearly slowed down in the late 1960s. Many energy techniques have reached unmistakable size and performance plateaus. This leveling-off is not a matter of technical limits but rather of prohibitive costs and unacceptable environmental impacts. Greater efficiency, reliability, and environmental compatibility have become new engineering goals.

Fossil fuels and electricity have been essential ingredients of far-reaching technical innovations in metallurgy, in the chemical industries, and in the development of new weapons. In metallurgy, important advances have included the large-scale production of inexpensive steel and aluminum. In the chemical industries the synthesis of ammonia and the production of numerous plastics constituted the greatest breakthroughs. Weapons with unprecedented destructive capabilities include high explosives, long-range guns, planes, armored warships, aircraft carriers, poisonous gases, missiles, and nuclear bombs.

But technical innovation has done surprisingly little to displace the established fuels. Coals and hydrocarbons are yielding only very slowly to innovations aiming to end the fossil-fuel era. Only nuclear electricity has made an important, but now faltering, contribution. Other nonfossil conversions appear decades away from gaining appreciable shares of global energy markets.

Fuels and Electricity

A temporary stoppage of the supply of fuel throws all the machinery of existence
out of action, and reveals the magnitude of the debt which civilized nations owe
to the men who win precious fuel from the earth's storehouse.
—Edward Cressy, *Discoveries of the Twentieth Century* (1930)

There have been two universal trends in coal production: the growing mechanization of underground extraction and increased surface mining. American productivities, the highest in the world, rose from less than 4 tonnes of coal per man per shift in 1900 to about 10 tonnes during the 1980s. In the world's largest surface mines, operating in the United States, Germany, Russia, and Australia, extraction has been as high as 30 tonnes per man per shift. The coal from such large mines has been increasingly burned in large adjacent power plants. When the coal must be transported to distant markets, it is loaded into special unit trains made up of up to 100 large, lightweight, permanently coupled hopper cars and pulled by powerful locomotives.

Advances in geophysical prospecting gave an immense boost to crude oil exploration. Rotary drilling, a technique used successfully for the first time at the Spindletop gusher in Beaumont, Texas, in 1901, sped up the development of new wells. In 1909 Howard Hughes introduced a device that assured the technique's dominance: the rolling cutter rock bit. By the 1930s the deepest oil wells surpassed 3000 meters, and they reached 9000 meters by 1980. Drilling has advanced steadily into more remote locations as well as into deeper offshore waters. The first well out of land sight was completed in 1947 off Louisiana. By the 1980s, jack-up and semisubmersible rigs were working in waters up to 2000 meters deep. The production platforms installed at major offshore fields are among the most massive structures ever built.

But the largest oilfields yet discovered have been on land, in the Persian Gulf region. The discoveries, which occurred between 1927 (Kirkuk in Iraq) and 1958 (Ahwaz in Iran), revealed vast supplies of oil. The large oil tankers developed after 1945 made it possible to ship this abundant fuel easily and inexpensively (see Figure 5.6; Ratcliffe 1985). Pipelines provide the most compact, most reliable, cleanest, safest mode of bulk transportation on land. The United States began to construct an extensive system of pipelines during the 1880s; the world's longest lines were built from western Siberia to Europe in the 1970s.

Two oil-refining processes—high-pressure cracking, introduced in 1913, and catalytical cracking, introduced in 1936—enabled the industry to produce lighter distillates from intermediate and heavy compounds. Only catalytic cracking can produce high-octane gasoline, the principal automotive fuel. Other advances, namely, large-diameter pipelines and turbine-driven compressors, made natural gas extraction feasible. The growth of natural gas industries was also a response to the rising demand for petrochemical products and clean household and industrial fuels. Dense networks of gas pipelines were built after 1945 in North America and after 1960 in Europe. The pipelines running from western Siberia to Europe, some as large as 2.4 meters in diameter, span a distance of nearly 6500 kilometers to convey a major portion of European supplies. For overseas deliveries, there are special tankers designed to transport liquefied natural gas. Such deliveries started in the 1960s, and Japan is now the largest importer.

Advancing electrification has required exponential increases in the ratings of all system components. The earliest small boilers were stoked with lump coal burned on moving grates. Starting in the 1920s, plant operators began to replace

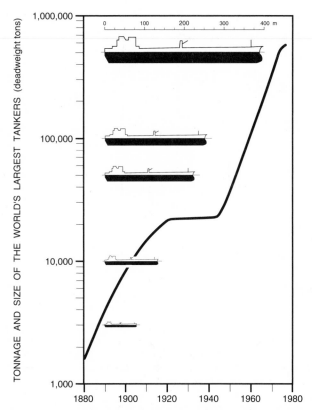

FIGURE 5.6 A century of oil tanker growth, from the *Gluckauf* (2307 deadweight tons), launched in 1884, to the *Seawise Giant* (564,709 deadweight tons), completed in 1976. *Source:* Scaled drawings assembled from data in various issues of *Tanker and Bulk Carrier.*

them with multistorey units that burned pulverized fuel injected into the combustion chamber. The boilers supply steam to turbogenerators, which themselves are much more efficient than the first units in operation a century ago. The modern turbogenerators operate with higher pressures and temperatures and their efficiencies may surpass 40 percent, compared to only 5 percent for the earliest machines (see Figure 5.7).

The expansion of utilities from urban to national systems has included the following universal components (Hughes 1983): the pursuit of economies of scale; the construction of larger stations in or near large cities; the development of high voltage links to transmit electricity from remote hydrostations; the promotion of mass consumption; and the interconnection of smaller systems in order to im-

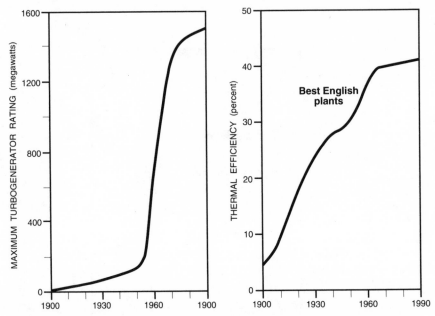

FIGURE 5.7 Power and thermal efficiency ratings of turbogenerators. Improvements in turbogenerators allowed them to operate with higher steam temperatures and pressures and to produce electricity with higher conversion efficiencies. *Sources:* Plotted from data in Dalby (1920), Federal Power Commission (1964), and various issues of *Mechanical Engineering* and *Power Engineering*.

prove supply security and to lower installed and reserve capacities. After 1950, responding to concerns about air pollution, utilities began to locate new generating stations close to fuel sources. This shift to mine-mouth power plants further increased the need for high-voltage transmission.

Consequently, the power of the largest transformers has grown 500 times and the highest transmission voltages have risen more than a hundredfold since the 1890s. Transmission started with wood poles and solid copper wires. Eventually it advanced to steel towers carrying steel-reinforced aluminum cables charged with up to 765 kilovolts (see Figure 5.8). Household service has also grown: The first houses with electricity had only a handful of sockets; today, houses commonly have complex wiring with more than fifty switches and outlets.

Some of the earliest generating stations tapped the power of falling water. With the advent of state-supported development in the United States (the New Deal's Tennessee Valley Authority) and the Soviet Union (part of Stalinist industrialization) during the 1930s, utilities were able to construct massive dams and install large turbogenerators. By 1990 hydrostations provided just over one-fifth of all electricity generation worldwide. Hydrostations supplied most of the electricity in many African and Latin American countries.

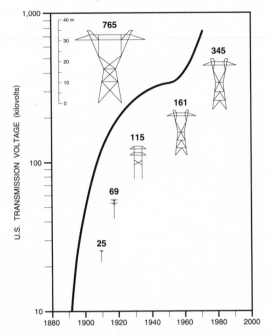

FIGURE 5.8 Evolution of U.S. transmission voltages. *Sources:* Plotted from data in Hughes (1983) and based on drawings in various issues of *Power Engineering.*

Nuclear electricity generation underwent very rapid commercial development. The first reactor delivered power in 1956, just fourteen years after the possibility of a sustained chain reaction was demonstrated on December 2, 1942. The pressurized water reactor (PWR), based on a design for U.S. nuclear submarine propulsion, became the dominant type worldwide. Although not a superior design, its early adoption and heavy military development made it entrenched by the time other reactors were ready to compete (Cowan 1990). By 1990, PWRs were installed in about four-fifths of all nuclear stations.

By 1990 nuclear fission generated about 17 percent of the world's electricity. France in particular embraced the new technique—there, nuclear power generated 70 percent of the national supply of electricity. Technical weaknesses, high construction costs, and the unresolved problem of long-term disposal of radioactive wastes have prevented the nuclear industry from making further inroads. Safety concerns and public perceptions of intolerable risks have always plagued the industry. The 1979 Three Mile Island accident and the 1986 Chernobyl disaster strengthened these concerns and refocused attention on alternative energies. But geothermal steam, direct solar radiation, and wind have found only limited applications thus far, and photovoltaic generation has only been essential in space applications where its high cost is not an obstacle.

Prime Movers

The history of the development of industrial society has been a history of plowing surplus energy back into more energy converters.

—**Earl Cook, Man, Energy, Society (1976)**

Twentieth-century engines range from small automobile motors to rockets that deliver missiles and space vehicles. The Otto cycle engine has developed rather conservatively. Its compression ratios have doubled, and its weight has declined. Its mass to power ratio fell from 40 grams per watt in 1890 to just around 1 gram per watt a century later. Its average power ratings increased steadily until the 1960s and then leveled off. America's first mass-produced car, Ransom Olds's Curved Dash, had a single-cylinder, 7-horsepower engine. Ford's Model T was three times as powerful; Ford produced 16 million units in the nineteen years from 1908 to 1927. Most cars on the road in the 1990s rate between 60 and 130 horsepower.

The internal combustion engine changed many agricultural practices. Engines began to power field machinery first in America; this change occurred shortly after the first success of passenger cars (Dieffenbach and Gray 1960). The first tractor factory was set up in 1905. Manufacturers introduced power take-off for attached implements in 1919 and power lift, diesel engines, and rubber tires between 1930 and 1932. The design trend has been toward lighter engines and frames. Modern machines require just one-fifth of the mass per unit of power output compared to the early designs and are up to six times more powerful. In Europe and Russia the widespread conversion to tractor power came only after 1950. The shift is still under way in most poor countries. In Asia's rice fields tractors are usually small two-wheeled machines. In 1990 the entire poor world had only about 10 percent more tractors than the United States alone, or just one-fifth of the global total of more than 26 million machines (FAO 1992).

Aircraft engines improved very rapidly after the Wright brothers' first flight (see Figure 5.9). The engines powering Boeing's 1936 *Clipper* were about 130 times more powerful than the Wrights' 1903 machine; the latter was ten times heavier than the *Clipper*'s engines per unit of power output (Gunston 1986). Diesels, too, got lighter and more powerful (Williams 1972). Modern diesels powering oil supertankers and other very large carriers are commonly more than seventy times more powerful than the engines deployed on World War I submarines. Similarly powerful engines have been installed to generate electricity in remote locations or to provide standby or emergency power. Like other types of engines, the diesels powering locomotives, earth-moving machinery, trucks, and buses have become lighter. The lightest diesels (just between 2 and 5 grams per watt) have powered passenger cars since the 1950s.

Gas turbines are a twentieth-century invention, a totally new prime mover that has revolutionized flying as well as many industries. The first gas turbine designs

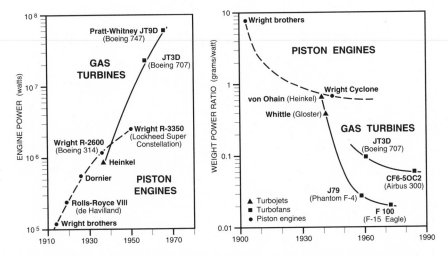

FIGURE 5.9 Evolution of aircraft engines. Increasingly more powerful yet lighter aircraft engines have assured progress in flying. Just before piston engines reached their limit of performance, gas turbines started their spectacular advance. Those powering large Boeing and Airbus planes now weigh less than 0.1 grams per watt, a hundredfold improvement compared to Wright's pioneering piston engine. Military jet engines are lighter still. *Sources:* Plotted from data in Constant (1981), Gunston (1986), and Taylor (1989).

emerged during the late 1930s when Frank Whittle in England and Hans Pabst von Ohain in Germany built experimental gas turbines for military planes (Constant 1981). Rapid development followed after 1945. The Bell X-1 airplane surpassed the speed of sound on October 14, 1947, and since then designers have introduced scores of supersonic fighter and bomber planes.

Britain built the first passenger jet, the Comet, in 1952. The jet failed, however, because of structural defects in its fuselage. Boeing introduced the first successful jet plane, the Boeing 707, in 1958. The wide-bodied Boeing 747 made its first commercial flight in 1969, and since then this jumbo plane has maintained its primacy on long-haul passenger flights (see Figure 5.10). The supersonic Concorde, which made its first commercial flight in 1976, has been expensive to operate and has not captured the market. Gas turbines have also found notable stationary applications, such as powering centrifugal compressors and electricity generators. The former are used in natural gas pipelines, in many chemical processes, and in the steel mills.

The only prime movers that can deliver more power per unit of weight than gas turbines are the rocket engines launching missiles and space vehicles. The old idea of rocket propulsion was turned into a mighty prime mover by developments started during World War II. The ethanol-powered engines of German V-2 missiles delivered 1 tonne of explosives to a target up to 340 kilometers away (von Braun and Ordway 1975).

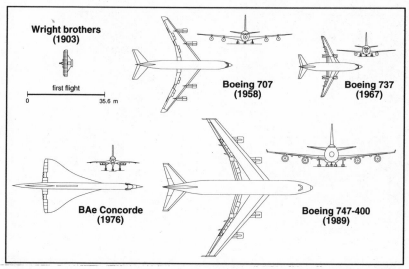

FIGURE 5.10 Plans and front views of some notable jet planes. The Boeing 707 (*top center*) was based on an inflight-refueling tanker. The Boeing 737 (*top right;* first flown in 1967) is the all-time best-selling jet aircraft. The supersonic French-British Concorde (*bottom left*) has remained an expensive oddity. The Boeing 747 (*bottom right*) was the first, and still is the most important, wide-bodied, long-haul aircraft. For comparison with these scaled drawings, the Wright brothers' plane and its total flight path on December 7, 1903, are shown (*top left*). *Sources:* Based on Boeing and Aerospatiale/BAe publications and Jakab (1990).

The space race between the superpowers began when the Soviets launched the Earth's first artificial satellite, the *Sputnik,* in 1957. As the race continued, the superpowers produced increasingly more powerful, and also more accurate, intercontinental ballistic missiles. On July 16, 1969, the eleven kerosene- and hydrogen-burning engines of America's Saturn C5 rocket started the Apollo spacecraft on its journey to the Moon. The engines were fired for just 150 seconds, and their combined power equaled that of more than 400 V-2s.

Metals

> *This age has been called the Iron Age, and it is true that iron is the material of most great novelties. By its strength, endurance, and wide range of qualities, this metal is fitted to be the fulcrum and lever of great works.*
>
> —W. Stanley Jevons, *The Coal Question* (1865)

Traditional metallurgy has undergone a huge expansion since the beginning of the fossil-fuel era. Metallurgists have mass produced new metals, developed new alloys, and devised new smelting methods. The coke-fueled smelting of iron

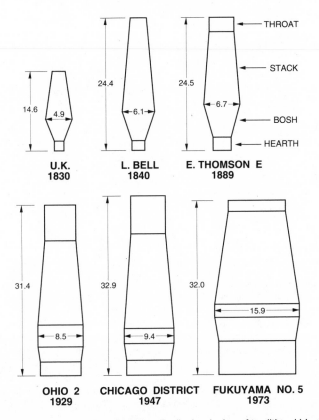

FIGURE 5.11 Evolution of blast furnaces. Radical redesign of traditional blast furnaces started with Lowthian Bell's work during the 1830s. Principal trends have included taller and wider stacks, larger hearths, and lower and steeper boshes. *Sources:* Based on Boylston (1936), King (1948), and Sugawara et al. (1976).

paved the way for other developments—and iron remains by far the most important metal. In 1990, the world's iron industry produced nearly thirteen times more pig (cast) iron than the combined total for five other leading metals: aluminum, copper, zinc, lead, and tin (UNO 1992). Iron production techniques have not changed dramatically since the invention of blast furnaces (see A5.2). Steel is the mainstay of most modern industrial and transportation infrastructures and an indispensable material in heavy construction.

The first important nineteenth-century innovation in iron-making was the use of the hot blast, patented by James Neilson in 1828. This technique saved large amounts of coke, as did a later innovation designed to recover and reuse the hot gases escaping from the top of the furnace. During the 1830s Lowthian Bell redesigned a typical furnace by making it both higher and wider (Bell 1884; see Figure 5.11). After 1870, Pennsylvanian ironmakers took the technical lead from Britain

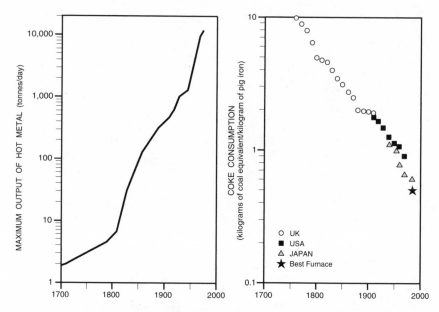

FIGURE 5.12 Semi-logarithmic graphs charting blast furnace efficiencies, 1700–1990. Internal furnace volume has increased from less than 20 to more than 5000 cubic meters and the maximum daily hot metal output has risen by almost four orders of magnitude since 1700 (*left*). Coke consumption rates in pig iron production have undergone almost perfectly exponential long-term declines. U.S. and Japanese technical improvements fit the trend established by British ironmakers before 1900. *Sources:* Calculated and plotted from data in Bell (1884), King (1948), Heal (1975), Sugawara et al. (1976), Dartnell (1978), and Gold et al. (1984).

and began working with much larger hearths and higher blast pressures (Temin 1964). Wartime demands during the first half of the twentieth century further strengthened America's iron-making primacy. After 1945, iron producers developed techniques for smelting under pressure and began to use highly beneficiated ores. They lowered their coke consumption even further by enriching the blast air with oxygen and injecting fuels (gas, oil, or pulverized coal) into the furnace (Gold et al. 1984).

By the 1960s Japan emerged as the most innovative ironmaker and built blast furnaces of unprecedented size and sophistication. Only the former Soviet Union surpassed Japan's total pig iron output. The Soviet blast furnaces were also large but not as efficient as the Japanese ones. Between 1800 and 1990 the global output of pig iron grew about 1500-fold, from less than 400,000 tonnes to more than half a billion tonnes a year. Since the introduction of coke, the internal volume of the largest furnaces grew more than 200 times. The highest daily output rose from around 3 tonnes to just over 12,000 tonnes (see Figure 5.12). The concurrent exponential decline in coke consumption reduced the total energy use per kilogram of metal by about 95 percent.

Large-scale steelmaking started with innovations introduced by Henry Bessemer in England after 1856. Bessemer produced steel in his pear-shaped tilting convertor by decarburizing molten pig iron with blowing cold air (Bessemer 1905). During the 1860s Siemens-Martin furnaces began to supplant Bessemer's convertor. Iron ore and iron scrap were added to the molten metal in shallow open hearths, and alternate blasts of gas made from coal and preheated air were blown over the charge (Schubert 1958). After 1879, steel's ascent was further accelerated by Thomas-Gilchrist basic brick linings, which removed the phosphorus from iron.

During the last quarter of the nineteenth century, steelmakers began to produce steel in huge quantities and in batches large enough to make outsize parts. Steel began to sustain industrial civilization in countless ways. During the last decades of the nineteenth century it was used to build growing rail networks, bigger ship hulls, longer bridges, and taller buildings. The first skyscraper rose in Chicago in 1888. In the twentieth century steel went into car bodies, turbogenerators, drilling rigs, pipes, and an enormous variety of tools and machines.

Open hearths dominated steelmaking until after 1950. The large-scale production of pure oxygen, developed in the 1930s, opened the way for the basic oxygen furnace. In this process, introduced in 1948 and rapidly adopted since the 1960s, molten pig and scrap iron lose their impurities by reacting with supersonic oxygen jets. The oxygen furnace uses no more than a quarter of the energy used by the open hearth. The energy costs of steel have been further lowered by continuous casting of the hot metal. This innovation supplanted the traditional production of ingots, which required reheating before further processing.

By far the most important innovation in nonferrous metallurgy was the development of aluminum smelting. The element was isolated in 1824, but an economical process for its large-scale production was devised only in 1886 when C. M. Hall in the United States and P.L.T. Heroult in France, working independently, electrolyzed aluminum oxide. Separating the metal requires more than six times more energy than smelting iron. Consequently, aluminum smelting advanced only slowly even after the beginnings of large-scale electricity generation. During the 1880s, smelting 1 tonne of the metal used more than 50,000 kilowatt hours of electricity. Steady improvements lowered this rate by more than two-thirds by 1990.

The aviation industry found new uses for aluminum. Metal bodies displaced wood and cloth during the late 1920s, and the demand rose sharply during World War II. After 1945 aluminum became a substitute for steel wherever the design required a combination of lightness and strength. These uses have ranged from automobiles to railway hopper cars to space vehicles. In turn, since the 1950s titanium has been replacing aluminum in high-temperature applications, above all in supersonic aircraft. Its production is at least three times more energy-intensive than the process for smelting aluminum.

Fertilizers and Pesticides

No single element of the indispensable materials is superior to any other, but all have equal value for the life of the plant. Therefore, if one element is missing from the soil, the others cannot produce a properly developed plant until the missing element has been supplied.

—Justus von Liebig, *Letters on Modern Agriculture* (1859)

Justus von Liebig's (1843) "law of the minimum" spelled out the nutritional needs of productive farming. Its formulation came before inexpensive fossil fuels and electricity made it possible to mass produce synthetic fertilizers. Before this development, crop yields were constrained by the recycling of organic wastes and the planting of legumes. The new fertilizers removed these constraints. First, John Bennett Lawes pioneered a process of treating phosphate rocks with diluted sulfuric acid to produce ordinary superphosphates. This method became common after 1870. Bennett's innovation spurred the discoveries of new phosphate deposits in Florida in 1888 and in Morocco in 1913. At the same time, potash mining began to expand slowly both in Europe and North America. But breaking the nitrogen barrier was by far the most important advance. Chilean nitrates, discovered in 1809, still provided the only inorganic options for nitrogen fertilization in the early 1890s. These imported nitrates were supplemented by ammonium sulfate recovered from coking ovens.

In the cyanamide process, coke reacted with lime to produce calcium carbide, which was then converted into calcium cyanamide. The process was introduced commercially in Germany in 1898, but it did not spread because it had high energy requirements and demanded large amounts of high-quality coke. Another innovation, introduced at the beginning of the twentieth century, produced nitrogen oxide in an electrical arc. This method, however, required a large supply of cheap electricity. Radical change came only with a brilliant invention, the Haber-Bosch process of ammonia synthesis.

The feasibility of this process was first demonstrated in 1913. The first practical use was to supply wartime Germany with explosives. The first synthetic nitrogen fertilizers were produced only after World War I, in the early 1920s. Because of economic setbacks during the 1930s and World War II, global ammonia synthesis stayed below 5 million tonnes until the late 1940s. By the late 1950s, more than one-third of American farmers still did not use any synthetic fertilizers (Schlebecker 1975). Afterwards, the use of synthetic fertilizers grew exponentially worldwide; the total applications of synthetic nitrogen compounds increased to nearly 85 million tonnes of nitrogen by 1990 (see Figure 5.13). Natural gas is the leading feedstock and now also supplies most of the energy for fertilizer synthesis.

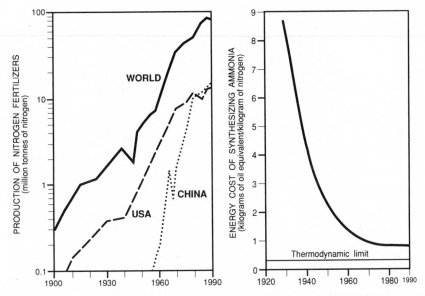

FIGURE 5.13 Nitrogenous fertilizer production, 1920–1990. The synthesis of nitrogenous fertilizers has increased worldwide as the energy costs of ammonia synthesis have declined. *Sources:* Plotted from data in Kirk and Othmer (1947) and various editions of FAO's *Fertilizer Yearbook*.

The energy cost of ammonia synthesis is now only about one-third of the pre–World War II average. Ammonia, a gas under atmospheric pressure, is the dominant nitrogenous fertilizer used in North America and increasingly also in Europe. Solid urea is the leading compound throughout the poor world. Although the energy cost of urea is higher than the energy cost of ammonia, urea is easily stored, moved, and applied. The post–World War II growth of fertilizer applications has been accompanied by an expanding use of pesticides. These chemicals reduce weed, insect, and fungal infestations of crops. The first commercial herbicide was 2,4-D. This growth-regulating substance, marketed in 1945, kills many broad-leaved plants without serious injury to crops. The first insecticide was DDT, released in 1944 (Friedman 1992). By 1990 more than 50,000 pesticides, mostly derived through energy-intensive processes using petrochemical feedstocks, have been registered for thousands of specific applications.

Weapons

Today, weapons have become our major industry. ... We dedicate our most sophisticated science and technology to the fashioning of new instruments of war. We burden our economy and overtax ourselves in the mass production of weapons.

—Ralph E. Lapp, *Arms Beyond Doubt* (1970)

Not surprisingly, many technical advances have been rapidly adopted for destructive uses. New fuels and new prime movers first increased the power and effectiveness of existing techniques; later they made it possible to design new classes of weapons capable of unprecedented reach, speed, and destruction. These efforts have culminated with the construction of enormous nuclear arsenals and with the deployment of intercontinental ballistic missiles capable of reaching any target on the Earth.

A comparison of weapons in the nineteenth and twentieth centuries is revealing. The two principal classes of weapons used during the American Civil War (1861–1865)—infantry muskets and 12-pound guns (both muzzle-loading and with smooth bores)—would have been quite familiar to veterans of the Napoleonic wars (Mitchell 1931). In contrast, among the weapons dominating the battlefields of the World War II in 1945—tanks, fighter and bomber planes, aircraft carriers, and submarines—only the last existed, and merely in experimental stages, during the 1890s!

The high explosives of the late nineteenth century laid the foundations for twentieth-century wars. These high explosives, in turn, had become possible with the invention of a new class of chemicals prepared through the nitration of such organic compounds as cellulose, glycerine, phenol, and toluene (Urbanski 1967). J.F.E. Schultze prepared nitrocellulose in 1865, and Alfred Nobel introduced dynamite in 1867. Reformulations suitable for military use became available during the 1880s (French Poudre B in 1884, Nobel's ballistite in 1888). The most powerful prenuclear explosive, cyclonite, was introduced by Hand Henning in 1899. Like gunpowder, high explosives are self-oxidizing, but they deliver a far more powerful blast, creating a shock wave. The detonation velocity of dynamite is nearly four times, and that of cyclonite more than six times, that of gunpowder. These new explosives were used in new gun shells, mines, torpedoes, and bombs.

The combination of better explosives and the ready availability and higher quality of steel brought greater power and longer range to field guns. Their effective range increased from less than 2 kilometers during the 1860s to over 30 kilometers by 1900. Long-range guns, heavy armor, and steam turbines enabled Britain to launch a new class of powerful battleships. The British *Dreadnought*, completed in 1906, pioneered this wave. Other notable pre–World War I destructive innovations included machine guns, submarines, and the first prototypes of military planes.

The horrible trench stalemates of World War I were sustained by massive deployment of heavy field guns, machine guns, and mortar launchers. Neither poisonous gases (first used in 1915) nor the first extensive use of fighter planes and tanks (1916, heavily only in 1918) could break the hold of that massive fire power deployed in frontal attacks (Liddell Hart 1954). The interwar years saw a rapid development of tanks and fighter and bomber planes. All-metal bodies replaced the traditional wood-canvas-wire construction, and the first purpose-built aircraft carriers came in 1922–1923. These weapons launched the aggression of World War

II. The early German successes were largely a matter of rapid tank-led penetrations. The surprise Japanese attack on Pearl Harbor on December 7, 1941, could be accomplished only with a large carrier force.

The same classes of weapons were essential in pushing back the Axis tide. First it was a combination of excellent fighter planes (Supermarine Spitfires and Hawker Hurricanes) and radar during the Battle of Britain in August and September 1940. Then came America's effective use of carrier planes (starting with the pivotal Battle of Midway in 1942) and the crushing Soviet tank superiority (model T-42) during the Red Army's westward thrust. The postwar arms race was spurred on by the development of jet propulsion, German ballistic missiles (the V-2 in 1944), and the first nuclear bombs. The first nuclear explosion, the Trinity test, came on July 11, 1945. Hiroshima was destroyed on August 6, 1945, Nagasaki three days later.

The development of nuclear bombs requires an enormous amount of energy, mostly for separating the fissile isotope uranium (Kesaris 1977). The postwar nuclear arms race between the United States and the former Soviet Union started with building more powerful fission bombs. Then came the construction of fusion bombs (first exploded in 1952 and 1953). These nations amassed arsenals amounting to more than 20,000 warheads. The development and deployment of delivery systems have also required an enormously energy-intensive effort. These systems now range from field guns to long-range bombers and from nuclear submarines to intercontinental ballistic missiles and cruise missiles. Almost one-tenth of all commercial energy used worldwide since 1940 was consumed in developing and amassing these weapons.

Unprecedented Power and Its Uses

Changes of like magnitude had never occurred so swiftly before.
—William T. Reid, *The Energy Explosion* (The Melchett Lecture 1969)

Global statistics show a sustained exponential growth in fossil-fuel production since the beginning of the nineteenth century (see Figure 5.14). Coal mining grew a 100-fold, from 10 million to 1 billion tonnes, between 1810 and 1910, and it reached nearly 5 billion tonnes by 1990. Crude oil extraction rose about 300-fold, from less than 10 million tonnes in the late 1880s to more than 3 billion tonnes a century later. During the same period natural gas production rose 1000-fold, from less than 2 billion to nearly 2 trillion cubic meters. The best way to appreciate the aggregate enormity of these energy flows is to compare them with traditional uses.

The best estimates show the worldwide total of biomass fuel consumption rising from around 700 million tonnes in 1700 to about 1.8 billion tonnes in 1990.

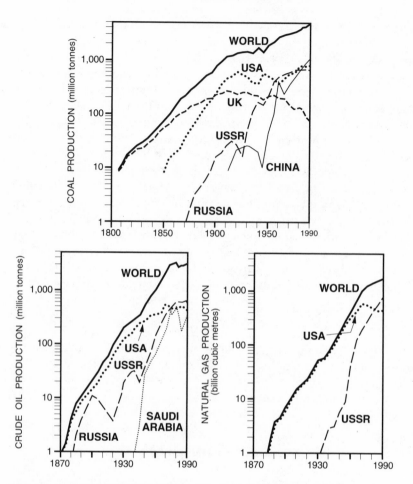

FIGURE 5.14 Global production of coal, crude oil, and natural gas. In each case, production has increased roughly a thousand-fold since the beginning of large-scale extraction. *Sources:* Plotted from data in UNO (1956) and various editions of UNO's *Energy Statistics Yearbook.*

This would be from about 250 million to 650 million tonnes in terms of oil equivalents (see Figure 5.15). During those nearly three centuries, extraction of fossil fuels rose from less than 10 million to the equivalent of almost 8000 million tonnes of oil. But the most meaningful comparison looks at this increase in terms of useful energy. Even in the early nineteenth century, coal combustion yielded more useful energy than the burning of biomass in open fires and fireplaces. The subsequent efficiencies of coal-fired boilers and stoves continued to rise. In terms of useful energy, the contributions of coal and biomass became equal sometime during the 1890s.

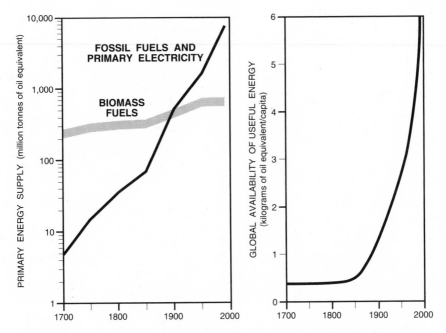

FIGURE 5.15 Global energy supply, 1700–1990. The worldwide output of fossil fuels surpassed the total supply of biomass energies just before the end of the nineteenth century, and it has grown since then more than tenfold (*left*). The global average of useful energy available per capita grew about twelvefold between 1850 and 1990 (*right*). *Sources:* Fossil fuel production plotted from data in UNO (1956) and various editions of UNO's *Energy Statistics Yearbook*; biomass consumption estimated by the author using data in Smil (1983).

Hydrocarbons and electricity are usually converted with even higher efficiencies than coal. Consequently, by 1950 commercial energies delivered about five times as much useful energy as biomass. By 1990 this difference grew to more than twenty times. In per capita terms this was more than twice the 1950 rate, five times the 1900 mean, and at least thirteen times the 1850 level. Global electricity output has been growing even faster. Less than 10 percent of all fossil fuels were converted to electricity in 1945, but the share reached a quarter in 1990. New hydro and nuclear capacities further expanded the generation of electricity.

The electricity supply was going up by 10.5 percent per year between 1900 and 1935 and by over 9 percent per year until 1970 (see Figure 5.16). In contrast, the annual exponential growth of fossil energies averaged only about 3 percent between 1900 and 1990. Since 1970 the growth of global electricity generation has been roughly halved. This decline reflects lower growth in demand throughout the rich world. The tenfold gain of total fuel use in less than 150 years stands in great contrast to the previous experience. For centuries, even for millennia, there was

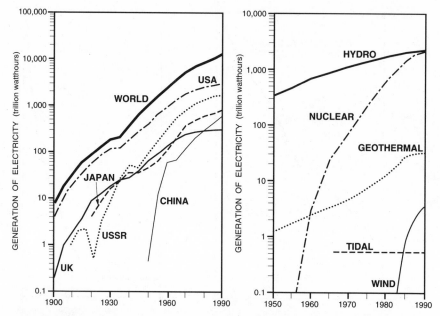

FIGURE 5.16 Generation of electricity, 1900–1990. Worldwide electricity generation has been growing considerably faster than the supply of fossil fuels. The largest economies have always been the leading producers, and thermal generation continues to dominate global output. By 1990 the share of electricity generated by nuclear power nearly equaled the share generated by water power. Geothermal and wind power remain marginal sources. Sources: Plotted from data in UNO (1956), U.S. Bureau of the Census (1975), Mitchell (1975), and various editions of UNO's *Energy Statistics Yearbook*.

hardly any change in per capita energy use in many traditional societies. Even the best-off regions saw only marginal increases. This impressive global rise has not been shared equally. The benefits of a high-energy society have gone disproportionately to a minority of the world's population.

Still, there have been many universal improvements and changes. No gain has been more fundamental than the substantial rise in global food production. No change has molded modern societies more than the process of industrialization. And no new developments have contributed more to the emergence of global civilization than the evolution of mass transportation and telecommunication.

Energy in Agriculture

A whole generation of citizens thought that the carrying capacity of the earth was proportional to the amount of land under cultivation and that higher efficiencies in using the energy of the sun had arrived. This is a sad hoax, for industrial man no longer eats potatoes made from solar energy, now he eats potatoes partly made of oil.

—Howard T. Odum, *Environment, Power, and Society* (1971)

Fossil fuels and electricity are essential inputs in modern farming. They are used directly to power machines and indirectly to build them, to extract and synthesize fertilizers and pesticides, and to develop new crop varieties. They have brought higher and more reliable yields, they displaced draft animals in all rich countries, and they greatly reduced their importance in the poor ones. The replacement of muscles by internal combustion engines and electric motors sustained the reduction of labor started by preindustrial farming advances. For example, average inputs to American wheat farming fell from about 30 hours per tonne of grain in 1800 to less than 2 hours per tonne by the late 1970s. This shift also led to huge declines in rural populations and to the worldwide rise of urbanization. A good early twentieth-century Western horse worked at a rate equal to the labor of at least six men—but even the early tractors had power equivalent to that of fifteen to twenty heavy horses.

American statistics illustrate the resulting displacements. The country's rural labor fell from more than 60 percent of the total workforce in 1850 to less than 40 percent in 1900. The share was 15 percent in 1950 and a mere 2 percent after 1975. Draft animals reached their highest numbers between 1910 and 1920. At that time their total power was about ten times as large as that of all the newly introduced tractors. But already by 1927 the two kinds of prime movers had equal power capacity, and the peak animal total was halved by 1940. By 1963, when America's tractor power was nearly twelve times the record draft animal capacity of 1920, the Department of Agriculture had stopped counting draft animals.

Until the 1950s mechanization proceeded much more slowly in Europe. In the populous countries of Asia and Latin America it really started only during the 1960s. By 1990 the American experience had been replicated in Canada and in the richest parts of Europe. Latin America's 1990 shares were generally over 20 percent, but in the three largest Asian economies—China, India, and Indonesia—two-thirds of the labor force were still farming (FAO 1992). Mechanization alone could not have released so much rural labor. Higher crop yields, brought by new crop varieties responding to higher fertilization and to more widespread irrigation, were also necessary.

Fertilizer use has been highest in Western Europe. During the 1980s the rates were generally over 300 kilograms of nitrogen per hectare, with appropriately smaller amounts of other nutrients. The top regional rates have been in the grainfields of northwestern Europe, from Denmark to France. Chinese rates averaged about 200 kilograms of nitrogen, and Japanese rates were about 150 (FAO 1990). The peak East Asian applications are in the Japanese rice paddies of Honshu and in China's coastal provinces of Jiangsu, Zhejiang, and Guangdong. Other regions of very high nitrogen use are Java, the Indian Punjab, the Nile Delta, and the American Corn Belt.

Synthetic nitrogen now supplies about half of the nutrient used annually by the world's crops. Because about three-quarters of all nitrogen in food proteins come from arable land, at least one-third of the protein in the current global food supply is derived from Haber-Bosch ammonia synthesis. Western nations, which use most of their grain as feed, could easily reduce their dependence on synthetic nitrogen by lowering their high meat consumption. Populous poor countries, where all but a small share of grain is eaten directly by humans, do not have that option. Most notably, synthetic nitrogen provides about 60 percent of all inputs in China. With over 80 percent of the country's protein supplied by crops, roughly half of all nitrogen in China's food comes from synthetic fertilizers (Smil 1992).

The global extent of irrigation grew fivefold between 1900 and 1990, from less than 50 million to more than 220 million hectares. This was a rise from less than 5 percent to about 15 percent of the world's harvested cropland (L'vovich et al. 1990). Half of this area is irrigated with pumped water, and about 70 percent of it is in Asia. Where farmers irrigate with water drawn from deeper aquifers, fuels or electricity for pumps are invariably the largest energy inputs in cropping. Most pumps still supply water to traditional ridges-and-furrows. More efficient, but also more expensive, sprinklers are used mostly in rich countries (Paul 1983).

A summary of global energy inputs to crop farming shows that in 1990 about half of the total went into the production of inorganic fertilizers (see A5.3). Almost three-fifths of these energy inputs were used in rich countries. But some poor populous nations already have become highly dependent on fossil energies for their food production. In the late 1980s China was using inputs costing nearly three times as much energy per hectare as in the United States, and Egypt's inputs were more than twice as much (Smil 1992). Other populous countries will soon follow this trend of rising dependence on fossil energies in food production. Productivity gains resulting from higher energy inputs have sustained an unprecedented growth of global population. More than that, they have also improved the typical nutrition for a rising share of the world's population.

Between 1900 and 1990 the world's cultivated area expanded by about one-third—but the global crop harvest rose nearly sixfold. This increase came about because of more than a fourfold rise in average crop yields; in turn, most of this advance must be attributed to an eighty-fold increase in energy inputs to field farming (see Figure 5.17). This achievement changed profoundly the global availability of food. In 1900 the gross global crop output (before storage and distribution losses) provided only a small margin above the average human food needs. Naturally, the share of the harvest that could be used to feed animals was very restricted.

In 1990 the global crop harvest allowed an average daily food availability twice as large as the normal need—for a population more than three times as large as in

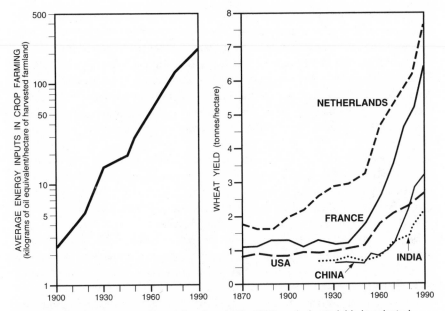

FIGURE 5.17 Energy inputs in crop farming, 1900–1990, and wheat yields in selected countries, 1870–1990. The world's cultivated land expanded only by about one-third between 1900 and 1990, but average energy inputs per cultivated hectare increased more than eighty-fold (*left*). Rising applications of fertilizers and pesticides and growing use of machinery and irrigation translated into much higher yields for all staple crops. *Sources:* Yield trends plotted from a variety of historical statistics and various editions of FAO's *Production Yearbook*.

1900! Consequently, plenty of grain could be fed to livestock. Rising consumption of meat and dairy products has brought high-protein diets to all rich nations and assured, on the average, adequate nutrition even in most of the world's largest poor populous countries. Most notably, the energy content of China's average diet has been less than 10 percent behind the Japanese diet since the mid-1980s (Smil 1986). Major improvements in Indian, Indonesian, and Mexican average food availability can also be traced to higher fuel and electricity inputs.

Industrialization

> *The provision of a whole new system of electric generation emancipated industry from the leather belt and line shaft, for it eventually became possible to provide each tool with its own electric motor. … Without high speed tools and the finer steels which they brought about, there could be nothing of what we call modern industry.*
>
> —Henry Ford and Samuel Crowther, *Edison as I Know Him* (1930)

Critical ingredients of the industrialization process include a large number of interconnected changes (Blumer 1990). The rising share of machine production concentrated in factories governed by hierarchical control required the location of workers near these establishments and the formation of new skills and occupations. The utilization of a money economy and the mobility of labor and capital established new contractual relations. Quests for mass output and low unit cost created new large markets. The rising availability of coal-derived heat and mechanical power produced by steam engines was not necessary to initiate these changes.

Cottage and workshop manufacturing, based on cheap countryside labor and serving national and even international markets, was going on for generations before the beginning of coal-energized industrialization (Mendels 1972; Clarkson 1985). This protoindustrialization had a considerable presence in parts of Europe (Ulster, Cotswolds, Picardy, Westphalia, Saxony, Silesia, and elsewhere). Voluminous artisanal production for domestic and export markets was also present in Ming and Qing China, in Tokugawa Japan, and in parts of India. Partially mechanized and relatively large-scale manufacturing of textiles based on water power was frequently the next step in the European transition. In a number of locations industrial waterwheels and turbines competed successfully with steam engines for decades after the introduction of the new prime mover.

Nor was mass consumption a real novelty. We tend to think of materialism as a consequence of industrialization, but in parts of Western Europe it was a major social force already during the fifteenth and sixteenth centuries (Mukerji 1981). The tastes and aspirations of increasing numbers of wealthier people provided an important cultural impetus to industrialization. They sought access to goods ranging from assortments of mundane cooking pots to exotic spices and fine textiles and from fascinating engraved maps to delicate tea services.

The term "industrial revolution" is as appealing and deeply entrenched as it is misleading. The process of industrialization was a matter of gradual, and often uneven, advances. This was the case even in regions that had moved rather rapidly from domestic manufactures to concentrated large-scale production for distant markets. The spuriously accurate timing of these changes (Rostow 1965) ignores the complexity and the truly evolutionary nature of the whole process. As already noted, the English beginnings of industrialization go back at least to the late sixteenth century. Its full development in Britain came only toward the middle of the nineteenth century (Clapham 1926; Ashton 1948). Even at that time traditional craftsmen greatly outnumbered machine-operating factory workers. According to the 1851 census the country still had more shoemakers than coal miners, more blacksmiths than ironworkers (Cameron 1985).

To view the worldwide industrialization process largely as imitative waves of English developments (Landes 1961) is no less misleading. Even Belgium, whose advances resembled most closely the British progress, followed a distinct path. There was a much greater stress on metallurgy and much lower concentration on textiles. Critical national peculiarities resulted in far from uniform industrialization patterns. They included the French emphasis on hydroenergy, America's and Russia's long-lasting reliance on wood, and Japan's tradition of meticulous craftsmanship. Coal and steam were initially no revolutionary inputs. Gradually they came to provide heat and mechanical power in unprecedented amounts, at high rates, and with great reliability. Industrialization could then be broadened and speeded up at the same time, eventually becoming synonymous with ever higher consumption of fossil energies.

Coal mining was not necessary to prepare a country for industrial expansion—but it was certainly critical for speeding it up. The contrast between Belgium and the Netherlands illustrates this effect. The highly urbanized Dutch society, equipped with excellent shipping and with relatively advanced commercial and financial capabilities, fell behind coal-rich, although otherwise poorer, Belgium, which became the most industrialized continental country in mid-nineteenth-century Europe (Mokyr 1976). Other European economies that took off early with coal-based industries included the Rhein-Ruhr region, Bohemia and Moravia within the Habsburg empire, and both Prussian and Austrian Silesia.

The pattern was repeated outside Western and Central Europe. In the United States, Pennsylvania, with its high-quality anthracites, and Ohio, with its excellent bituminous coal, emerged as the early leaders (Eavenson 1942). In pre–World War I Russia it was the discovery of rich Donets coal deposits in the Ukraine and the development of Baku oilfields during the 1870s that ushered in the subsequent rapid industrial expansion (Falkus 1972). Japan's quest for modernity during the Meiji era was energized by coal from northern Kyushu. This coal fueled the country's first blast furnace at Yawata ironworks, the predecessor of Nippon Steel, beginning in 1901. India's largest commercial empire grew from J. Tata's blast furnace using Bihari coke in Jamshedpur starting in 1903.

Once energized by coal and steam power, traditional manufactures could turn out larger volumes of good-quality products at lower prices. This achievement was a necessary precondition for mass consumption. Availability of an inexpensive and reliable supply of mechanical energy also allowed increasingly sophisticated machining. In turn, this led to more complex designs and greater specialization in manufacturing of parts, tools, and machines. New industries energized by coal, coke, and steam were set up to supply national and international markets with unprecedented speed. High-pressure boilers and pipes started to be manufactured after 1810. The production of rails and railway locomotives and wagons rose rapidly after 1830, as did the making of water turbines and screw propellers after 1840. Iron hulls and submarine telegraph cables found new large markets after 1850, and Bessemer and Siemens-Martin steel processes found wide acceptance after 1860.

Rising fuel inputs and the replacement of tools by machines reduced human muscles to a marginal source of energy. Labor increasingly became responsible for supporting, controlling, and managing the productive process. Still, new systematic studies of individual tasks and complete factory processes demonstrated that labor productivity could be greatly increased by optimizing, rearranging, and standardizing muscular activities (Taylor 1911). This approach had to start with detailed studies of labor sequences. Often it led to an accelerated pace of work and higher demands on the labor force, but it also brought many labor-sparing and environment-enhancing improvements.

A radically new period of industrialization came when steam engines were rapidly eclipsed by electrification. Electricity is a superior form of energy not only in comparison with steam power. Only electricity combines instant and effortless access with the ability to serve very reliably every consuming sector except for flying. The flip of a switch converts it into light, heat, motion, or chemical potential. Its easily adjustable flow allows for previously unsustainable precision, speed, and process control. Moreover, it is clean and silent at the point of consumption. And once a proper wiring is in place, electricity can accommodate an almost infinite number of growing or changing uses—yet it requires no inventory.

These attributes made electrification of industries a truly revolutionary switch. After all, steam engines replacing waterwheels did not change the way of transmitting mechanical energy powering various industrial tasks. Consequently, this substitution did little to affect general factory layout. Space under factory ceilings remained crowded with mainline shafts linked to parallel countershafts transferring the motion by belts to individual machines. The outage of a prime mover (whether caused by low water or an engine failure) or a transmission failure (caused by a line shaft crack or a slipped belt) disabled the whole setup. Such arrangements also generated large frictional losses and allowed only limited control of power at individual workplaces.

The first electric motors powered shorter shafts for smaller groups of machines. After 1900 unit drives rapidly became the norm. In American manufacturing this transformation was virtually complete in just three decades. Between 1899 and 1929, the capacities of industrial electrical motors grew from less than 5 percent to over 80 percent of all installed mechanical power (Devine 1983; Schurr 1984). This efficient and reliable unit power supply did much more than remove the overhead clutter with its inevitable noise and risk of accidents. The demise of the shaft drive freed the ceilings for installing superior illumination and ventilation and made possible a flexible plant design and easy capacity expansion. Highly efficient electric motors—combined with precise, flexible, and individual power control in a better working environment—brought much higher labor productivities.

Electrification also launched vast specialized industries. First came the manufacturing of light bulbs, dynamos, and transmission wires (after 1880) and steam and water turbines (after 1890). High-pressure boilers burning pulverized fuel were introduced after 1920, and giant prestressed concrete dams began to be constructed a decade later. Widespread installation of air pollution controls came after 1950, and the first nuclear power plants were commissioned before 1960. The rising demand for electricity has also stimulated geophysical exploration and fuel extraction and transportation. A great deal of fundamental research in material properties, control engineering, system design, and automation was also necessary.

The availability of reliable and cheap electricity has transformed virtually every industrial activity. By far the most important effect on manufacturing was the widespread adoption of assembly lines. Their classical, and now outdated, rigid Fordian variety was based on a moving conveyor introduced in 1913. Modern, flexible Japanese procedures rely on just-in-time delivery of parts and on skilled workers capable of doing a number of different tasks. Electricity has energized the smelting of aluminum, and it has been indispensable in making plastics and composite materials. These lightweight materials have been widely substituted for steel. Without electricity there could be no large-scale micromachining to produce parts for such exacting applications as jet engines or medical diagnostic devices. And, of course, there would be neither any accurate electronic controls nor the omnipresent computers.

Although the manufacturing shares have been steadily declining in virtually all rich countries, the industrialization process continues. But instead of running on massive inputs of energy and human capital, new industrial production is sustained by growing inputs of information and services. Research, design, marketing, and servicing have become as important as the actual making of goods. Quality considerations, rather than preoccupation with quantity, dominate this process. This trend has major implications both for future energy use and for the structure of the labor force. Given its recent rapid advances, this so-called second industrial revolution may eventually deserve its name.

Transportation

> Until the early 19th century, transport had altered little for 2000 years. Then, quite suddenly, the whole situation changed with the advent of railways. ... This was indeed a transport revolution, but the 20th century was to see two further revolutions of comparable magnitude, first on the roads and, later, in the air.
>
> —Trevor I. Williams, *The History of Invention* (1987)

Several attributes apply to all forms of fossil-fueled, or electrified, transport. In contrast to traditional ways of moving people and goods they are not only much faster, often almost incredibly so. They are also incomparably more reliable, substantially cheaper, and capable of transferring at one time much larger numbers of passengers or much greater masses of goods.

For millennia no transportation was faster than riding a good horse. For centuries no conveyance was less tiring than a well-sprung coach. Railways did away with these constants in a matter of years. They not only shrank distances and redefined space but did so with unprecedented comfort. The speed of a mile a minute (96 kilometers/hour) was first reached on a scheduled English run in 1847. By coincidence this was also the year of the greatest railway-building activity in the country, which laid down a dense network of reliable links within just two generations (O'Brien 1983).

Very soon passenger cars ceased to be merely carriages on rails and acquired heating and washrooms. For a higher price there was also good upholstering, fine meal services, and sleeping arrangements. Faster and more comfortable trains carried not only visitors and migrants to cities, but also urbanites to the countryside. Thomas Cook was already offering railway holiday packages in 1841. Commuter lines started the first great wave of suburbanization. Increasingly capacious freight trains brought bulky resources to distant industries and speedily distributed their products.

The total length of British railways was soon surpassed by American construction, which was begun in 1834 in Philadelphia. By 1860 the United States had 48,000 kilometers of track, three times the U.K. total. By 1900 the difference was nearly tenfold. The first transcontinental link came in 1869, and by the end of the century there were four more such lines (Haney 1968). Russian development also progressed fairly rapidly. Less than 2000 kilometers of track were laid by 1860, but the total rose to over 30,000 by 1890, and to nearly 70,000 kilometers by 1913 (Falkus 1972). The transcontinental link across Siberia to Vladivostok, begun in 1891, was not fully completed until 1917. When the British withdrew from India in 1947 they left behind about 54,000 kilometers of track (and 69,000 in the whole subcontinent). No other mainland Asian country built a major railway network before World War II (Saxena 1962).

Since the end of World War II, competition by cars, buses, and planes has reduced the relative importance of railways in most industrialized countries. The most successful innovations have been fast long-distance electrical trains. Japanese *shinkansen,* first run in 1964, reaches a maximum of 250 kilometers per hour. French *trains à grand vitesse,* operating since 1983, can go up to 297 kilometers per hour. Post–World War II Russia and some industrializing countries (Brazil, Iraq, Algeria) have also been vigorous builders of new lines. China has been the Asian leader, with over 30,000 kilometers of track added between 1950 and 1990.

The first steamships crossed the North Atlantic no faster than the best contemporary sailing ships with favorable winds. But already by the late 1840s the superiority of steam was clear, with the shortest crossing time cut to less than ten days (see Figure 5.18). By 1890 trips of less than six days were the norm, as were steel hulls. Steel did away with size restrictions: Structural considerations limited the length of wooden hulls to about 100 meters. The large ships of such famous lines as Cunard, Collins, or Hamburg-America became proud symbols of the technical age. They were equipped with powerful engines and double-screw propellers, furnished with grand staterooms, and offered excellent service.

The opulence of these great liners contrasted with the crowding, smells, and tedium of steerage passages, especially on smaller ships. By 1890 steamships carried

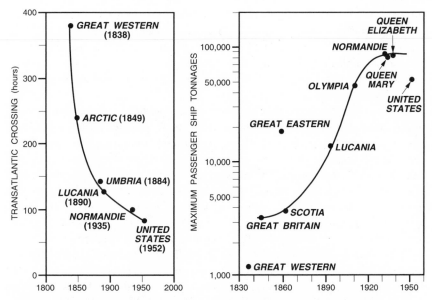

FIGURE 5.18 Transatlantic crossing times, 1800–1960, and maximum passenger ship tonnages, 1830–1950. As the ships traveling between Europe and North America grew in size and acquired more powerful prime movers (steam engines until the 1890s, steam turbines afterwards), the time needed for crossing the Atlantic was cut from more than two weeks to just over three days. *Sources:* Plotted from data in Fry (1896), Croil (1898), and Ayres (1969).

more than half a million passengers a year to New York. By the late 1920s the total North Atlantic traffic surpassed 1 million. But by 1957, five years after the launch of the fastest passenger vessel, airlines carried more people between Europe and North America than ships. A decade later, regularly scheduled transatlantic service came to the end.

The world's fleet of commercial steamships got its early boosts from the completion of Suez Canal (1869) and from the introduction of effective refrigeration (during the 1880s). Its later growth has been stimulated by the opening of the Panama Canal (1914), deployment of large diesel engines (after 1920), and transport of crude oil. Since the 1950s larger specialized ships have been needed not only to move oil but also widely traded bulky commodities (above all ores, lumber, grain, and chemicals) and growing shipments of cars, machinery, and consumer goods.

Changes brought by private cars—economic, social, and environmental—rank among the most profound transformations of the modern era (Flower and Jones 1981; Ling 1990; Womack et al. 1991). In country after country (for the first time in the United States during the mid-1920s), car making emerged as the leading industry in terms of product value. Cars have also become major commodities of international trade. Their exports from Germany (after 1960) and even more so from Japan (after 1970) have clearly strengthened those two economies. Large segments of other industries—above all steel, rubber, glass, plastics, and oil refining—became vitally dependent on making and using cars. Highway building has commonly involved massive state participation leading to enormous cumulative capital investments. Hitler's *Autobahnen* of the 1930s preceded Eisenhower's system of interstates by a generation.

Certainly the most obvious car-generated change has been the reordering of urban space. Worldwide marks of this transformation include the proliferation of freeways and parking spaces and the destruction of neighborhoods. Where space has allowed, there has also been a rapid increase in suburbanization (in North America, also in exurbanization) and changes in the locations and forms of shopping and services. But the social impacts have been even greater. Car ownership has been an important part of *embourgeoisement.* Affordable family classics enjoyed amazing longevity (Siuru 1989). The first one was Ford's Model T, whose price dropped as low as $265 in 1923. Other notable models were the Austin Seven, the Morris Minor, the Citroen 2CV, the Renault 4CV, the Fiat Topolino and, the most popular of them all, Ferdinand Porsche's Hitler-inspired Volkswagen.

The freedom of personal travel that these cars provided to millions of families has had enormous effects on both residential and professional mobility. These benefits have proved to be highly addictive. Kenneth Boulding's (1974) analogy of car as a mechanical steed turning its driver into a knight with an aristocrat's mobility looking down at pedestrian peasants is hardly exaggerated. In 1990 there were only 1.75 people per car in the United States, and the rates were 1.9 in Germany and 2.1 in Japan (MVMA 1992). This widespread addiction makes it difficult for rich countries to give up the habit (see Figure 5.19). They have gone to extraordinary lengths to preserve this privilege. They have also been willing to put up with enormous death, injury, and pollution costs. Western nations should not be surprised that Chinese and Indians want to emulate this habit.

Trucking has also had many profound socioeconomic consequences. Its first mass diffusion—in rural America after 1920—reduced the cost and speeded up the movement of farm products to market. These benefits have been replicated first in Europe and in Japan, and during the past two decades also in many Latin American and Asian countries. In rich countries long-distance heavy trucking has become the backbone of food deliveries as well as a key link in the distribution of industrial parts and manufactured goods. In many rapidly growing economies it has obviated the construction of railways and opened up remote areas to commerce and development—but also to environmental destruction. In most poor nations buses have become the leading means of long-distance passenger transport.

Scheduled international air transport started with daily London-Paris flights in 1919 and advanced to regular transoceanic links just before World War II. The age of mass air travel came only with the introduction of jet aircraft in the late 1950s. By 1990 three major manufacturers (Boeing and McDonnell Douglas in America

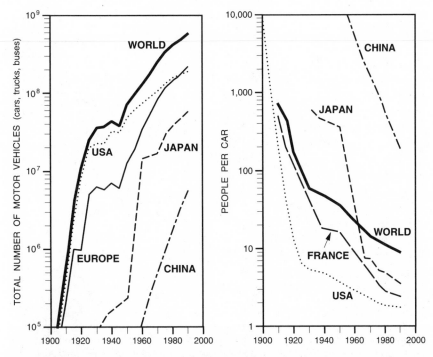

FIGURE 5.19 Motor vehicles, 1900–1990. The number of cars, trucks, and buses in use worldwide grew from about 10,000 in 1900 to nearly 600 million in 1990. U.S. registrations were surpassed in the late 1980s by the European total, but the country still has the highest rate of ownership, with just 1.75 people per car in 1990. The global rate dipped to below 10 people per car by the mid-1980s (the United States had reached this level by the early 1920s). *Source:* Plotted from materials compiled by the Motor Vehicle Manufacturer's Association.

and the Airbus consortium in Europe) were building about 30 different types of planes. About 20 of these were wide-bodied aircraft capable of flying up to nearly 500 people.

The speed and range of these planes (see Figure 5.20), the proliferation of airlines and flights, and the nearly universal linking of reservation systems has made it possible to travel among virtually all major cities of the planet in a single day. Naturally, cargo deliveries can duplicate this feat. Moreover, the costs of flying have been steadily declining in real terms, in part because of lowered fuel consumption. These achievements opened up new business opportunities as well as mass long-distance tourism to major cities and to subtropical and tropical beaches. They also opened up new possibilities for unprecedented migrant and refugee movements, for widespread drug smuggling, and for international terrorism involving aircraft hijacking.

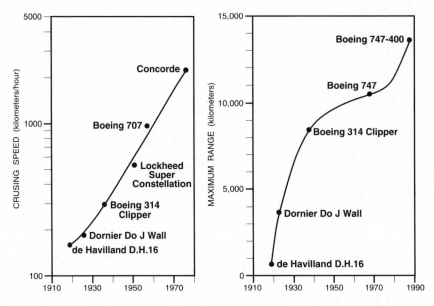

FIGURE 5.20 Aircraft cruising speeds and maximum ranges, 1910–1990. The first scheduled commercial flights (on the de Havilland D.H. 16 in 1919) averaged just over 150 kilometers per hour for a maximum range of about 600 km. By the late 1950s, Boeings could cruise at close to 1000 km/h, and by the late 1980s the Boeing 747–400 could fly 13,500 km nonstop. The Concorde, flying at over twice the speed of sound, has been a costly exception rather than a precursor of a new generation of planes. *Source:* Plotted from data in Taylor (1989).

Information and Communication

If we consider information transfer in its broadest sense, the information industry,
with all its complex ramifications, now employs more people than any other.

—Trevor I. Williams, *The History of Invention* (1987)

Fossil-fueled societies produce, store, distribute, and use incomparably larger amounts of information than their predecessors. They do so in a still growing variety of forms whose capacities show as yet few signs of reaching practical limits. Except for printing, all of these techniques were developed during the high-energy age. Except for photography and early phonographs none of them could function without electricity.

In East Asia and in early modern Europe printing was a well-established commercial activity for hundreds of years before the introduction of fossil fuels (Johnson 1973). But hand typesetting was laborious and print runs were limited by the slowness of hand-operated wooden screw presses. Iron frames speeded up the work to more than 200 sheets per hour. Only steam-operated machines, introduced by Frederick Koenig in 1810, made it possible to produce mass editions rapidly. By the 1820s runs of 2000 sheets per hour were possible, and rotary presses raised this rate tenfold by the late 1850s. The displacement of hand composition by mechanical typesetting started during the 1880s. Linotype machines could cast 5000 to 7000 pieces of type per hour, monotypes up to 12,000. Photocomposition techniques introduced after 1950 increased these rates eventually to more than 1 million characters per hour.

Inexpensive and reliable telecommunication became possible only with electricity. The first century of its development was dominated by messages transmitted by wires. Decades of experiments in various countries ended with the first practical telegraph, introduced by William Cooke and Charles Wheatstone, in 1837. Its success depended on a reliable source of electricity provided by Alessandro Volta's battery, designed in 1800. Adoption of the coding system of Samuel Morse in 1838, and the rapid extension of land lines in conjunction with railways, were the most notable early developments.

Undersea links (across the Channel in 1851, across the Atlantic in 1866) and a wealth of technical innovations (including some of Edison's early inventions) combined to make telegraphs global within just two generations. By 1900 multiplex wires with automatic coding carried millions of words every day. The messages ranged from personal to diplomatic codes and included reams of stock market quotations and business orders.

The telephone, patented by Alexander Graham Bell in 1876 just hours ahead of Elisha Gray's independent filing (Hounshell 1981), had even faster acceptance for local and regional service. Reliable and cheap long-distance links were introduced rather slowly. The first trans-American link came only in 1915, and the transatlantic telephone cable was laid only in 1956. To be sure, radio-telephone links were available from the late 1920s, but they were neither cheap nor reliable.

The storage, reproduction, and transmission of sound and pictures was developing concurrently with telephone advances. Thomas Edison's 1877 phonograph was a simple hand-operated machine, as was Emile Berliner's more complex gramophone in 1888. Electric record players took over only during the 1920s. Image-making advanced rather slowly from its French beginnings in the 1820s and 1830s (J. N. Niepce, L.J.M. Daguerre). Developments speeded up after 1890 with breakthroughs in cinematography (the first public showing by the Lumiere brothers in 1895) and the introduction of color film. Sound movies came in the late 1920s and the invention of xerography a decade later (by Chester Carlson in 1937).

The quest for wireless transmission started with Heinrich Hertz's generation of electromagnetic waves in 1887. Subsequent practical progress was fast. By 1899 Guglielmo Marconi's signals crossed the Channel, two years later the Atlantic. In 1897 Ferdinand Braun invented the cathode-ray tube, the device that made possible both television cameras and receivers. In 1906 Lee de Forest built the first triode. Its indispensability for broadcasting, long-distance telephone service, and

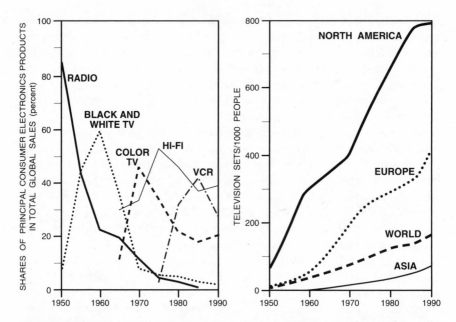

FIGURE 5.21 Shares of principal electronic products in total global sales. The first graph shows the distinct waves in post–World War II consumer demand for electronics. For an increasing number of people around the world, watching television has become the third most time-consuming commitment (after work and sleep). America is saturated with TV sets, but Asia still has a long way to go and Africa has barely started. *Sources:* Based on compilations prepared by the Japanese Ministry of International Trade and Technology and on data in various editions of UNESCO's *Statistical Yearbook*.

computers ended only with the invention of the transistor. Regular radio broadcasts started in 1920. The BBC offered the first scheduled television service in 1936, and RCA followed suit in 1939.

Post–World War II advances of information and communication techniques have been unprecedented both in number and in their innovative reach (Taylor 1982). The world of images was first enriched by instant pictures (Edwin Land's Polaroid in 1946) and by wide-screen projection. Then came the mass production of high-quality Japanese cameras (after 1960) and video-recorders (1970s) and new electronic modes of photography. After 1950 transistors downsized both radios and TVs, making them accessible worldwide. The consumer electronics industry benefited from the switch from black-and-white to color TV as well as from the demand for a variety of audiophile equipment. Long-playing records were introduced in 1948, stereo recordings a decade later. Phillips pioneered cassette tapes during the 1960s and introduced compact discs during the early 1980s. Mass buying of VCRs started at the same time (see Figure 5.21).

The era of accessible intercontinental calls came with automatic dialing via geostationary satellites. This innovation was made possible by a combination of microelectronic advances and powerful rocket launchers during the 1960s. Satellite broadcasting, begun extensively during the 1970s, has formed the basis of a truly global information system.

Enormous progress has also been made since the 1960s in designing and deploying a wide range of diagnostic, measurement, and remote sensing techniques. These advances have yielded a previously unimaginable wealth of information. X-rays, discovered by W. K. Roentgen in 1895, were the only such option in 1900. By 1990 these techniques ranged from radioactive imaging and ultrasound (used both in medical diagnoses and in engineering) to radar (developed on the eve of World War II). Since the early 1960s, satellite sensors, capable of acquiring data in various bands of electromagnetic spectra, have improved weather forecasting and natural resource management.

But the greatest post-1945 advances have been in the storage and processing of both numbers and words. By far the most important ingredient in building modern information societies has been the rapid development of integrated circuits placed on a single silicon chip. The great tradition of nineteenth-century mechanical calculators—starting with prescient designs by Charles Babbage after 1820 (Swade 1991)—was finally left behind with the development of the first electronic computers during World War II. The calculating speed of post–World War II programmable machines started to rise exponentially as transistors supplanted vacuum tubes (see Figure 5.22). Computers became both powerful and widely affordable only with the rapid progress in miniaturizing transistor circuits.

Starting with a single planar transistor in 1959, the number of components per chip doubled every year until 1972. Since that year it has continued to increase at only a slightly lower rate. This trend has brought rapidly declining costs creating huge new markets beyond personal computers. Microchips have found new applications especially in consumer electronics, telecommunication, precise manufacturing, and process controls.

A reliable supply of electricity is the critical precondition for all microelectronic devices—but their current profusion does not put a great strain on overall deliveries. The power ratings of these devices are very low in comparison with both their predecessors and with other common electrical gadgets (see A5.4). The microchip revolution, transforming not only economic but also social relations (Dertouzos 1991), is thus clearly a part of a broader trend toward a conservative use of high-quality energy.

Economic Growth

The flow of energy should be the primary concern of economics.
—Frederick Soddy, *Wealth, Virtual Wealth, and Debt* (1933)

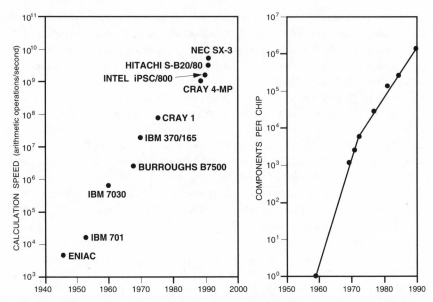

FIGURE 5.22 Advances in computers, 1946–1990. Between 1946 and 1990 the fastest computer speeds rose from 5000 to 5.5 billion operations per second, an increase of seven orders of magnitude representing twenty successive doublings of the highest capacity (left). Between 1959 and 1990 miniaturization techniques increased the density of transistor circuits on a silicon chip from 1 to more than 1 million (*right*). These developments led to the emergence, rising power, and falling cost of personal computers. *Sources:* Plotted from data in Ralston (1976), Moreau (1984), Meindl (1987), Chen (1990), and Corcoran (1991).

Advances in agricultural and industrial production and the growing importance of transportation and information flows have combined to sustain unprecedented increases in the total economic output of every modernizing country. This rise has been modified by the cyclical nature of economic expansion (van Duijn 1983) and interrupted by major internal or international conflicts. But both its magnitude and its cumulative effects have been impressive (Kuznets 1971; Rostow 1978). Traditional economies were either largely stationary or managed to grow by a few percentage points per decade. In contrast, the industrializing societies of the nineteenth century sustained decadal expansion rates of between 20 and 60 percent. Such growth rates meant that the output of the British economy in 1900 was nearly ten times larger than in 1800. America's gross domestic product (GDP) doubled in just twenty years between 1880 and 1900. Japanese output during the Meiji era (1868–1912) rose 2.5 times.

Even with rather high population increases these performances translated to large per capita GDP gains. They ranged mostly between 15 and 20 percent per decade in Western Europe and North America and averaged about 25 percent in Japan. After the setbacks of the first half of the twentieth century (the two world wars and a global economic recession in the 1930s), worldwide economic growth resumed at even faster rates. There had never been a period of such rapid and widespread growth of output and prosperity as between 1950 and 1973, before OPEC quintupled the world price of crude oil. The American per capita GDP, already the world's highest, rose by 60 percent. The West German rate more than tripled, and the Japanese one more than sextupled. A number of poor populous countries of Asia and Latin America also entered a phase of vigorous economic growth. In spite of record population increases, China and Mexico managed to double their per capita GNPs. Brazil did even better.

Conversions of national outputs to a common denominator (usually US dollars) are needed to summarize the total world economic product. These values almost invariably distort the real purchasing power of currencies. Consequently, aggregates of the gross world product (GWP) are not the best indicators of real growth—but they at least capture the relative magnitude of past increases. In terms of constant 1990 US dollars the GWP grew from about $1 trillion in 1900 to nearly $4 trillion in 1950. In 1973 it was almost $14 trillion, a nearly fourfold increase in a single generation!

The steady decline of crude oil prices before 1970 was a critical ingredient of this unprecedented expansion. OPEC's first round of oil price increases, which began in 1971 and ended in a fivefold rise after the Yom Kippur War of October 1973, temporarily stopped this growth. The second round of oil price raises was ushered in in 1979 by the overthrow of the Iranian monarchy and the ascent of fundamentalist *shia* clerics to power. The global economic slowdown of the early 1980s was accompanied by record inflation and high unemployment.

The long-term economic importance of rising energy consumption is also indisputable. Obviously, both the initiation and the maintenance of strong economic growth are matters of complex, interdependent inputs. They are unthinkable without intellectual advances ranging from better management to vigorous inventing and patenting. They require technical improvements and responsive institutional arrangements, most notably sound banking and legal systems. Appropriate government policies, good education systems, and a high level of competitiveness are also essential. But none of these inputs would have made much difference without the concurrently rising consumption of fuels and electricity.

This link is made clear by long-term national trends (see Figure 5.23) as well as by global energy-economy rankings. When national per capita means of GDP are adjusted for purchasing power parities they show a very high correlation with average energy consumption rates. But the growth of absolute energy consumption with higher levels of economic development hides an important relative decline. Maturing economies tend to have lower energy intensity: They consume diminishing amounts of fossil fuels per unit of GDP. This reality reflects the combination of the declining importance of energy-intensive capital inputs, improved conversion efficiencies, and the rising shares of the service sector (see Figure 5.24).

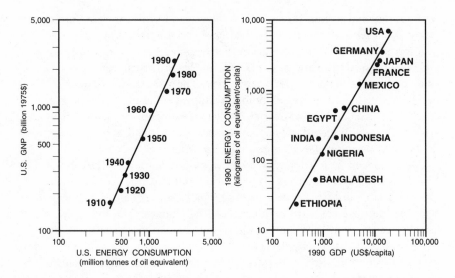

FIGURE 5.23 U.S. energy consumption, 1910–1990, and energy consumption of selected countries, 1990. The close correlation between the growth of a nation's primary energy consumption and its expanding economic product is shown by the U.S. record during the twentieth century (*left*). The global comparison indicates a relationship between energy use and stages of economic growth (*right*). Sources: Plotted from data in U.S. Bureau of the Census (1975), Summers and Heston (1991), and various editions of UNO's *Energy Statistics Yearbook*.

The British peak came around 1850, and the U.S. economy had its highest energy intensity around 1920. Japanese energy intensity peaked only in 1970, and the Chinese maximum was reached during the late 1970s. Although the total energy intensity of the world economy has been declining since the 1920s, the consumption of electricity per unit of economic product is still growing. In 1990 it was more than twice as high as in 1950, but it may be peaking. Indeed, in some industrialized countries the measure declined slightly for the first time during the 1980s.

Consequences and Concerns

No machine is an abstract force moving through history. Rather, every new technology is a social construction and the terms of its adoption are culturally determined.

—David E. Nye, *Electrifying America* (1992)

Extensive and often stunning technical advances made possible by the unprecedented use of fossil fuels and electricity have created a mixture of benefits and

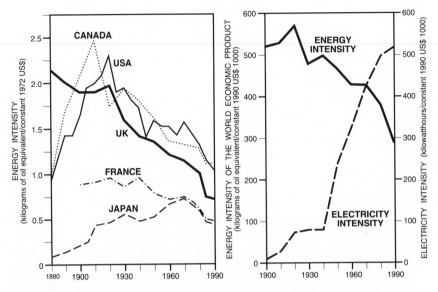

FIGURE 5.24 Energy intensity of selected economies, 1880–1990, and energy intensity of the world economic product, 1900–1990. The declining energy intensity of the gross domestic product has been a universal feature of maturing economies. This trend also clearly applies to the world's total economic product. In contrast, electricity inputs per unit of the gross world product are still increasing. *Sources:* Calculated from data in various editions of UNO's *Energy Statistics Yearbook*, OECD's *Economic Indicators*, and the CIA's *Handbook of Economic Statistics*.

problems. On close examination even the most desirable advances turn out to have considerable drawbacks. Unanticipated social costs and environmental penalties have been the most worrisome problems. The still continuing worldwide process of urbanization is a perfect illustration of these complexities. So is, of course, the rise of mass consumption, a process underlying major improvements in the quality of life. The Western, or Japanese, combination of urbanization and relative affluence holds an enormous appeal to hundreds of millions of families on the three poor continents—but it rests on clearly unsustainable foundations.

The economic imperatives of securing reliable energy supplies have had a number of undesirable political consequences. On one hand they have led to policies of uncontrolled resource development and subsidized waste, on the other they have engendered rigid regulation and monopolistic inefficiency. Internationally, they have dictated desirable quests for free trade, cooperation, and foreign aid—but they also have played an important role in framing and justifying aggression and in directing the course of armed conflicts. And even should the world succeed in preventing any thermonuclear exchange, it still faces a growing concern about the future of high-energy civilization because of the environmental effects of growing fossil-fuel use.

Urbanization

*Urbanized societies, in which a majority of the people live crowded together in
towns and cities, represent a new and fundamental step in man's social evolution.
... Neither the recency nor the speed of this evolutionary development is widely
appreciated.*

— **Kingsley Davis**, *Scientific American* (1965)

Cities, even large cities, have been around for a long time (Chandler 1987). Rome
of the first century A.D. housed more than half a million people. Harun ar-
Rashid's early ninth century A.D. Baghdad had 700,000 people, and the Tang dynasty Changan of the same period counted around 800,000. A thousand years
later Beijing, the capital of Qing, topped 1 million, and in 1800 there were about
fifty cities worldwide exceeding 100,000. But even in Europe urban totals in 1800
were no more than one-tenth of the population. Subsequent rapid increases in
both the population of the world's largest cities and in the overall shares of urban
populations would have been impossible without fossil fuels.

Traditional (solar) societies could support only a small number of large cities
because the relatively high power densities of urban food and fuel consumption
had to be supplied by harvesting biomass energy from large surrounding areas.
Total food and fuel consumption of large preindustrial cities had to come from
croplands and woodlands at least 40 times, and commonly about 100 times, larger
than the size of the settlement itself (see A5.5). Moreover, absence of powerful
prime movers put clear limits on the capacity of transport of food and material
both to and within cities, as well as on deliveries of water and on removal of
waste.

Traditional cities had to be supported by the concentration of diffuse energy
flows, but fossil-fueled cities are supplied by the diffusion of concentrated energies. Better crop yields reduce farmland needs even with diets richer in animal
foods. The coalfields and oilfields that supply fossil-fueled cities may occupy an
area equal to no more than 10 percent and as little as one-tenth of 1 percent of
their built-up areas (see A5.5). New, powerful prime movers transport fuels from
their basically punctiform places of extraction to urban users spread over large areas. They also make large-scale, long-distance food imports affordable. And it is
simply unimaginable how cities with millions of people could pump their drink-
ing water, remove their sewage and garbage, and meet their transportation and
communication needs without fossil fuels and electricity.

The rise of fossil-fueled cities was rapid. In 1800, only one of the world's ten
largest cities—London (then number two)—was located in a country fueled by
coal. A century later, nine out of ten of the largest cities were in that category:
London, New York, Paris, Berlin, Chicago, Vienna, St. Petersburg, Philadelphia,
and Manchester. The only exception was Tokyo, in still largely biomass-fueled Japan. In 1900, cities still claimed only about 15 percent of the world's population—

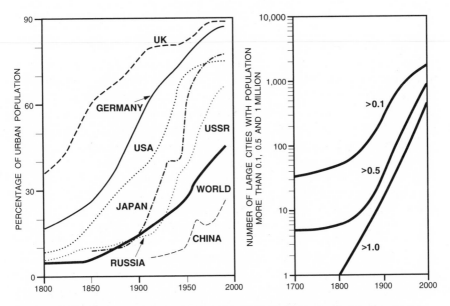

FIGURE 5.25 Urban population statistics, 1800–1990. By 1990 urbanization brought the global share of city populations to about 45 percent, roughly a tenfold increase in two centuries. In most rich countries the process has reached, or approached, plateaus at above 70 percent (*left*). The rapid increase in the number of very large cities still does not show any signs of saturation (*right*). Source: Plotted from data in Chandler (1987).

but the urban share of population was far higher in the world's three largest coal-producing nations. The rate was over 70 percent in the U.K., approaching 50 percent in Germany, and nearly 40 percent in the United States (see Figure 5.25). The subsequent continuation of urban growth has also brought a remarkable increase in the total number of very large cities. In 1990 more than 300 urban areas surpassed 1 million inhabitants, compared to 13 in 1900 and 1 in 1800.

Urban growth has been driven by the push of agricultural mechanization and the pull of industrialization. Urbanization and industrialization are not, of course, synonymous, but the two processes have been closely tied by many mutually amplifying links. Most notably, technical innovation in Europe and North America had overwhelmingly urban origins (Bairoch 1991). The massive shift of urban jobs into services is largely a post–World War II development. These transfers brought the urban share of total population up to more than 70 percent in most Western nations by 1990. The rapid post-1950 migration to Latin American cities pushed their share to nearly 70 percent of the continent's population by 1990. The industrialization of the former Soviet Union caused the same rise in the Russian urban population. Only African and Asian urban shares remained less than 50 percent of the population on the respective continents. Asia's low figure has been heavily influenced by decades of tightly controlled migration in Maoist China.

The economic, environmental, and social effects of these great human translocations have been among the most avidly studied phenomena of modern history. The misery, deprivation, filth, and disease common in rapidly growing nineteenth-century cities brought forth a vast literature. These writings ranged from primarily descriptive (Kay 1832) to largely indignant (Engels 1887). Similar realities—minus the threat of most contagious diseases eliminated by inoculation—can be seen today in many Asian, African, and Latin American cities. But people are still moving in. Then as now they are often leaving conditions that, on balance, were even worse. This fact was commonly neglected both by the authors of the original reformist writings and by those engaging in subsequent debates about the disadvantages of urbanization (Williamson 1982).

One must weigh the dismal state of urban environments—esthetic affronts, air and water pollution, noise, crowding—against their often no less objectionable rural counterparts. Common rural disamenities have included very high concentrations of indoor air pollutants from unvented biomass combustion, unsafe water supplies, and dilapidated, overcrowded housing. Moreover, the drudgery of field labor is seldom preferable even to unskilled industrial work. In general, typical factory tasks require lower energy expenditures than do common farm tasks.

In a surprisingly short time after the beginning of mass urban industrial employment, the duration of factory work became reasonably regulated. Later came progressively higher wages in combination with such benefits as health insurance and pension plans. Together with better educational opportunities these changes led to appreciable improvements of typical standards of living. This led eventually to the emergence of a substantial urban middle class in all largely *laissez faire* economies. The appeal of this great, although now certainly tarnished, Western accomplishment is felt strongly throughout the industrializing world. And it was undoubtedly an important factor in the demise of Communist regimes that proved slow to deliver similar benefits.

Standard of Living

> It is hard to think of an idea more immediate than that of the living standard. It
> figures a good deal in everyday thought. It is, in fact, one of the few economic
> concepts that is not commonly greeted with the uncommon scepticism reserved for
> other concepts of economics.
>
> —Amartya Sen, *The Standard of Living* (1987)

Acting via industrialization and urbanization, growing rates of energy consumption have been exerting fairly orderly, and largely desirable, effects on standards of living. In the early stages of economic growth these benefits are limited because fuels and electricity are overwhelmingly channeled into building up an industrial base. A slowly increasing consumption of household and personal goods—better cookware, clothes, and furniture—is the first sign of improvement. Since the 1970s the list has also included mass-produced electronic gadgets. Afterwards, along with better food supply and improving health care, come lower infant mortality and longer life expectancy.

Later in the process these basic material and health advantages start spilling into the countryside. The educational level of urban populations begins to rise, and there are increasing signs of incipient affluence. Eventually comes the stage of mass consumption with its many physical comforts and frequent ostentatious displays. Longer periods of schooling, high personal mobility, and growing expenditures on leisure and health are part of this change. This sequence unmistakably correlates with rising average per capita energy consumption rates. National peculiarities (from climatic to economic singularities) preclude any simple classification, but three basic categories are evident.

No country whose annual primary commercial energy consumption averages less than the equivalent of 100 kilograms of oil per capita can guarantee even basic necessities to all of its inhabitants. Bangladesh and Ethiopia are the most populous nations in this category in the 1990s. China belonged there before 1950, as did large parts of Western Europe before 1800. As the rate of energy consumption approaches the equivalent of 1 tonne of oil, industrialization advances, incomes rise, and quality of life noticeably improves. China of the 1980s, Japan of the 1930s and the 1950s, and Western Europe and the United States between 1870 and 1890 are outstanding examples of this group. Widespread affluence requires, even with efficient energy use, the equivalent of at least 2 tonnes of oil per capita per year. France made it during the 1960s, Japan during the 1970s.

The French example illustrates the speed of this change. The census of 1954 revealed the striking primitivity of housing in France. Less than 60 percent of the households had running water, only one-quarter had an indoor toilet, and only one in ten had a bathroom and central heating (Prost 1991). By the mid-1970s, refrigerators were in almost 90 percent of the households, toilets in 75 percent, bathrooms in 70 percent, and central heating and washing machines in about 60 percent. By 1990 all of these possessions became virtually universal. Three-quarters of all families also owned a car in 1990, compared to less than one-third in 1960. Such growing affluence was reflected in the rising use of energy. Between 1950 and 1960, per capita energy consumption rose by about 25 percent, but between 1960 and 1974 it soared by over 80 percent. Between 1950 and 1990 the per capita supply of all fuels more than doubled; during the same period, gasoline consumption rose nearly sixfold and electricity use went up more than eightfold.

The easing of women's household work in the Western world has been similarly recent. For generations rising fuel consumption made little difference for everyday housekeeping. Indeed, it could make it worse. As the standards of hygiene and social expectations rose with better education, women's work in Western countries often got harder. No matter if it was washing, cooking, and cleaning in cramped English apartments (Spring Rice 1939) or doing daily chores in American farmhouses, women's work was still exceedingly hard during the 1930s. Electricity was the eventual liberator. Regardless of the availability of other energy forms, without electrical appliances this form of labor remained exhausting and often dangerous (Caro 1982; see A5.6).

Many electric appliances were available already by 1900. General Electric was selling electric irons, vacuums, immersion water heaters, and cookers. The high cost of these appliances, limited house wiring, and slow progress in rural electrification delayed their widespread adoption, both in Europe and in North America, until the 1930s. Refrigeration has been a more important innovation than gas or electric cooking (Pentzer 1966). The first home refrigerators were marketed by the Kelvinator Company in 1914. American refrigerator ownership rose sharply only during the 1940s, and these appliances became really common in Europe only after 1960. Their importance has increased with the growing reliance on fast food. Refrigeration now accounts for up to 10 percent of all the electricity used in the households of rich nations.

Electricity continues to bring further time and labor savings to rich countries. Self-cleaning ovens, food processors, and microwave ovens (developed in 1945, but introduced in small household models only during the late 1960s) have become common throughout the rich world. Although, in recent years, the better-off segments of Asian and Latin American populations have been acquiring refrigerators, washing machines, and microwave ovens, hundreds of millions of women throughout the poor world are still awaiting this energy-driven liberation.

Although air conditioning was first patented by William Carrier in 1902, it was limited for decades to industrial applications. The first units scaled down to household use came during the 1950s. Their widespread adoption opened up the American Sunbelt to mass migration from northern states and has increased the appeal of subtropical and tropical tourist destinations. By 1990 household air conditioners were widely used not only in rich countries but also in the urban areas of the industrializing world.

Since 1950 a number of poor countries have become intermediate energy consumers. Still, in terms of total population, the distribution of global energy use remains extremely skewed. In 1950 only about 250 million people (one-tenth of the global population) consumed annually the equivalent of more than 2 tonnes of oil per capita—yet they claimed 60 percent of the world's primary energy (excluding biomass). By 1990 nearly one-quarter of the world's population consumed this much energy and this segment claimed nearly three-quarters of all fossil fuels and electricity. In contrast, the poorest quarter of humanity used less than 5 percent of all commercial energies, a disparity resulting in a roughly twenty-fold difference in average per capita consumption.

Stunning as they are, these averages do not capture the real difference in living standards. Poor countries devote a much smaller share of their total energy consumption to private household and transportation uses—typically less than 20 percent, compared to more than 30 percent for rich nations. The actual difference in typical direct per capita energy use among the richest and poorest quarters of humankind is thus closer to being forty-fold rather than "just" twenty-fold. This enormous disparity is the key reason for the chronic gap in economic achievements and quality of life. In turn, these inequalities are a major source of persistent global political instability.

Political Implications

The strategic importance of oil for military operations has become so great that it is impossible to envisage wars, conducted with today's weapons, without oil. Sociopolitically, oil consumption represents a yard-stick of economic prosperity and the quality of life.
—Awni S. Al-Ani, in *The Challenge of Energy* (1981)

The dependence of modern societies on massive, reliable, and inexpensive supplies of fossil fuels and electricity has generated a multitude of political concerns and responses, domestic and foreign. Perhaps the most universal concern is the concentration of decision-making power. As Richard Adams (1975) noted, when "more energetic processes and forms enter a society, control over them becomes disproportionately concentrated in the hands of a few, so that fewer *independent* decisions are responsible for greater releases of energy."

The greatest peril arises when these already concentrated controls become superconcentrated in a single individual. Their misdirection can result in enormous human suffering, prodigious waste of labor and resources, and destruction of cultural heritage. The unchallenged decisions of the two Communist dictators who converted their manias into terrible realities are the epitomes of this danger.

In 1953, the year of Stalin's death, the Soviet Union used more than twenty-five times more energy than it had in 1921, the year the country emerged from its civil war (Clarke 1972). Yet the generalissimo's paranoia led to the *gulag* empire and to the economic prostration of the world's potentially richest nation. Similarly, on Mao Zedong's death in 1976 China's energy production was more than twenty times the 1949 total (Smil 1988). But the Great Helmsman's delusions brought the Great Leap Forward followed by the worst famine in human history—between 1959 and 1961 nearly 30 million Chinese died (Ashton et al. 1984)—and the destruction of the Cultural Revolution.

Given the importance of crude oil in modern economies and the concentration of the richest hydrocarbon resources, the decisions of a few individuals in the Middle East can have profound consequences for global prosperity. During the early 1970s, a handful of members of the Saudi royal family and the shah of Iran together controlled more than one-third of the world's oil reserves. Their decision to increase oil earnings led to the quintupling of world oil prices in 1973–1974 and ushered in a period of high inflation and reduced economic growth. In response, all major Western importers and Japan set up energy-sharing agreements coordinated by the International Energy Agency. Individual countries also promoted closer bilateral ties with OPEC nations and subsidized the quest for domestic fuel self-sufficiency or for alternative sources of energy. France's development of nuclear electricity and Japan's energy conservation effort have been especially remarkable.

The concentration of energy production has a number of other—more mundane, but no less far-reaching—political consequences. In many market economies, direct state ownership of utilities, or at least close government supervision, were put in place to ensure a secure supply of electricity and provide for the often high capital needs of its development. Governments often manipulate energy prices for political purposes. Some governments subsidize essential energy supplies to buy the support of major constituencies; others institute special taxes on energy use to raise additional revenue. The first category includes Egypt, where fuel prices are below the world level, and China, which fixed coal prices even below the cost of production. The second category includes all the countries that heavily tax gasoline, that is, all rich nations outside of North America.

The failed but initially apparently successful attempt of the Arab OPEC nations to turn oil into a political weapon was not the first instance of energy supply carrying an ideological message. American industrialists displayed the power of light for the first time during the 1894 Columbian Exposition in Chicago and flooded the downtown areas of large cities with "White Ways" (Nye 1992). The Nazis used walls of light to awe the participants at party rallies in the 1930s (Speer 1970). Electrification became the embodiment of such disparate political ideals as Lenin's quest for a new state form and Roosevelt's New Deal. Lenin summarized his goal in a terse slogan: "Communism equals the Soviet power plus electrification." Roosevelt used extensive federal involvement in building dams and electrifying the countryside as a means of economic recovery (Lilienthal 1944).

The most important political problems of energy use today are not just national but global: the wide energy gap between rich and poor nations and the highly uneven distribution of crude oil reserves. Less than 2 percent of the world's oilfields, concentrated in just five of almost 300 hydrocarbon basins, store about three-quarters of global reserves (Perrodon 1985). The basins are, in ascending order, Lake Maracaibo in Venezuela, the Volga-Ural region, the Gulf of Mexico, Western Siberia, and the Persian Gulf.

The riches of the Persian Gulf are unparalleled: The basin has twelve of the world's fifteen largest oilfields and two-thirds of the world's oil reserves. The Saudi al-Ghawar oilfield, discovered in 1948, contains 15 percent of all known crude oil reserves. These riches explain the lasting interest of major oil importers in the region's stability. The near-chronic disarray of the area, made up of imposed states separated by arbitrary borders and ancient ethnic and religious enmities, complicates the situation immensely.

Outside intervention into Middle Eastern affairs in the post–World War II era started with the Soviet attempt to take over northern Iran (1945–1946). American troops landed twice in Lebanon (1958, 1982). Western countries made extensive arms sales and shipments to Iran (before 1979, during the last decade of Shah Reza Pahlevi's reign) and Saudi Arabia, and the Soviets did the same for Egypt, Syria, and Iraq. Western tilt (weapons, intelligence, and credit) benefited Iraq during the Iraq-Iran war (1980–1988).

Of course, the pattern of intervention culminated in the Desert Shield/Storm operation of 1990–1991, a massive U.S.-led, UN-sanctioned response to the Iraqi invasion of Kuwait. The Iraqi advance had doubled the oil reserves under Iraq's control, pushing them to about 20 percent of the global total, and seriously threatened the nearby Saudi oilfields and perhaps even the very existence of the Saudi monarchy, which controls one-quarter of the world's oil reserves.

Historically, resource-related objectives have not usually determined broader strategic aims (Lesser 1991). The quest for assured access to oil, however, was a critical factor in some of the twentieth century's most momentous military moves. Most notably, it had a proximate role in the Japanese decision to start the war in the Pacific in December 1941 (Sagan 1988). A year later it led to the diversion of the Nazi armies toward the Caspian oilfields, a move causing flank overextension and weakening the thrust at Stalingrad (Liddell Hart 1970).

Warfare

Nuclear weapons are not weapons of war. The only purpose they can possibly have is to deter their use by the other side, and for that purpose far fewer are good enough.

—Victor F. Weisskopf, *Bulletin of the Atomic Scientists* (1985)

The weapons of the twentieth century have demanded vast economic resources and incorporated enormous destructive power. For example, in August 1914 Britain had just 154 airplanes in all of its military forces—but by the summer of 1918 the country's aircraft factories were employing no fewer than 350,000 people, turning out planes at a rate of 30,000 units per year (Taylor 1989). World War II raised the mobilization of national economies for war to new highs. American industries delivered just 514 aircraft to the U.S. forces during the last quarter of 1940. Rapid industrial mobilization resulted in the total wartime production of over 250,000 planes, more than the combined German and British output (Holley 1964).

During World War II, the major combatants devoted huge resources to military spending. In 1944, the United States spent 54 percent of its net national economic product on the war, and the Soviet Union and Germany spent 76 percent of their net products on the war in 1942 and 1943, respectively (Harrison 1988). Ultimately, Allied victories were due to their superiority in harnessing destructive energy. By 1944, the United States, the Soviet Union, the U.K., and Canada were producing three times as many combat munitions as Germany and Japan (Goldsmith 1946). Deployment of this fire power reached incredible intensities. The most concentrated tank attack of World War I, on August 8, 1918, involved almost 600 machines. In contrast, nearly 8000 tanks were engaged during the final Soviet assault on Berlin in late April 1945. These tanks were supported by 11,000 planes and more than 50,000 guns and rocket launchers (Ziemke 1968).

The increasing destructiveness of weapons can be illustrated through several comparisons. Combat casualties during the Battle of the Somme (July 1–November 19, 1916) totaled 1.043 million. Those during the Battle of Stalingrad (August 23, 1942–February 2, 1943) surpassed 2.1 million (Craig 1973). Battle deaths—expressed as fatalities per 1000 men of armed forces fielded at the beginning of a conflict—stayed below 200 during the first two modern wars involving major powers (the Crimean War of 1853–1856 and the Franco-Prussian War of 1870–1871). They surpassed 1500 during World War I and 2000 during World War II—and more than 4000 for the Soviet Union during World War II (Singer and Small 1972). Germany lost about 27,000 combatants per million people during World War I and more than 44,000 during World War II.

Civilian casualties increased at an even faster rate. During World War II some 40 million civilian casualties accounted for more than 70 percent of the 55 million total. Large cities suffered huge losses within days or just hours of bombings (Kloss 1963). In Germany, the number of bombing casualties reached nearly 600,000 dead and almost 900,000 wounded. In Japan, about 100,000 people died during nighttime raids by B-29 bombers, which leveled about 83 square kilometers of four principal Japanese cities between March 10 and March 20, 1945. Five months later two nuclear bombs killed at least 100,000 people (see A5.7).

Paradoxically, nations with huge stockpiles of thermonuclear weapons have not used them mainly because they are so much more destructive than conventional weapons. If the United States and the Soviet Union had engaged in a limited thermonuclear exchange targeting strategic facilities rather than cities during the late 1980s, the direct effects of blast, fire, and ionizing radiation would have caused at least 27 million and up to 59 million deaths (von Hippel et al. 1988). Such a prospect has acted as a very powerful deterrent against launching or, since the 1960s, even seriously contemplating a first attack.

Environmental Changes

Thus human beings are now carrying out a large scale geophysical experiment of a kind that could not have happened in the past nor be reproduced in the future.

—Roger Revelle and Hans Suess, in *Tellus* (1957)

Fossil fuel energy consumption is the largest cause of anthropogenic air pollution and a leading contributor to water pollution and land use changes. Coal and liquid fuel combustion releases a huge amount of particulate matter, sulfur, and nitrogen oxides. Water pollution arises mainly from accidental oil spills, refinery operations, and acid mine drainage. And major land use changes are caused by coal mining, reservoirs created by major dams, right-of-way corridors for high-voltage transmission lines, and extensive storage, refining, and distribution facilities for liquid fuels.

Indirectly, fuels and electricity are responsible for many other pollution flows and ecosystemic degradations. The most notable ones arise from industrial production (above all from metallurgy and chemical syntheses), agricultural chemicals, urbanization, and transportation. Although these impacts have been with the industrializing world for generations, they have been increasing both in extent and intensity (Turner et al. 1990). All major economies have had to devote growing attention to environmental management. Scientists now suspect that these effects may compromise the long-term viability of major biospheric functions and bring undesirable changes on the global scale (Smil 1993).

Among a number of such worries only a few phenomena have a truly global reach. The dispersion of long-lived chlorinated pesticides and radioactive fallout from atmospheric nuclear testing were among the first concerns of this kind. Concentration of the stratospheric ozone protecting the planet from excessive ultraviolet radiation has been declining, and the Earth's changing radiation balance may lead to rapid climatic changes. The first process was accurately predicted in 1974 and has been acutely demonstrated above the Antarctic since 1985 (Rowland 1989). Ozone loss has been caused largely by releases of chlorofluorocarbons (CFCs). International agreements adopted since 1987 aim at an early replacement of CFCs by less harmful compounds. No comparable technical fix is available to deal with the large-scale generation of greenhouse gases, which are altering the radiation properties of the atmosphere.

The leading contributor is carbon dioxide, the end product of efficient combustion of all fossil and biomass fuels. Since the mid-nineteenth century the global generation of this gas has risen exponentially with the increasing consumption of fossil fuels. Atmospheric concentrations of the gas rose by almost 40 percent between 1850 and 1990. Conversions of grasslands and forests also contributed to the rise. Other greenhouse gases include methane, nitrous oxide, and CFCs (see A5.8). These atmospheric changes could lead to a relatively rapid global rise of surface temperatures (Houghton et al. 1990). The complexity of the involved atmospheric, oceanic, and biospheric processes precludes any reliable predictions about the rate of the possible change. But there is no doubt that high-energy civilization has been engaged in an unprecedented geophysical experiment on a planetary scale.

This process could have enormous social implications (Committee on Science, Engineering, and Public Policy 1991). Rapid global warming (an increase of just 1–

4 °C in less than a century) would have a wide range of undesirable consequences. They could include new precipitation patterns, coastal flooding, shifts in ecosystem boundaries, and the spread of warm-climate, vector-borne diseases. Changes in plant productivity, loss of near-shore real estate, sectoral unemployment, and large-scale migration from affected regions would be the key economic conse- quences. The only potentially successful way to deal with these changes is through unprecedented international cooperation. This worrisome challenge also offers motivation for making fundamental changes in managing human affairs.

APPENDIXES

A5.1 Fossil Fuels

Fossil fuels span a huge range of properties and qualities. Solid fuels range from nearly pure carbon in anthracites to lignites and peats of low-energy density. Bituminous (black) coals make up the bulk of global extraction. Because they nearly always contain significant shares of ash and sulfur, their combustion generates fly ash (rather easy to remove from hot gases) and sulfur dioxide (expensive to capture and hence mostly released to the air). Crude oils are mixtures of complex hydrocarbons containing only traces of ash. Their re- fining yields a variety of products for specific final uses: gasolines, jet and diesel fuel for transportation, fuel oils for heat and steam generation, and also lubricants and paving ma- terials. Natural gases, the cleanest fuels to burn, are simpler hydrocarbons dominated by methane.

Hydrocarbons can also be produced from the conversion of coal. Low-energy town gas was widely used for public and domestic lighting during the nineteenth century. Modern coal gasification can produce synthetic gas, which is equal to natural gas in terms of qual- ity. Synthetic liquid fuels were produced for the first time on a large scale by Germany dur- ing World War II.

Energy densities vary widely in coals but are relatively uniform for hydrocarbons. Crude oils are always superior: They contain nearly twice as much energy per unit mass as com- mon bituminous coals. International energy statistics use one of three common denomi- nators: standard coal equivalent, a fuel containing 29.3 MJ/kg (7 Mcal/kg); oil equivalent of 42 MJ/kg; or values in standard energy units (calories, joules, or British thermal units).

About 80 percent of the world's coal resources are in Russia, the United States, and China, while about two-thirds of all oil resources are in the Persian Gulf region. The world's coal resources could last for almost 500 years at the 1990 rate of extraction. Hydro- carbon resources are smaller: The median value is about 100 years for the eventual produc- tion of all estimated oil resources and just over 200 years for natural gas (White 1987).

Fossil Fuel	Energy Density	
	MJ/kg	*MJ/m³*
Coals		
Anthracites	31–33	
Bituminous coals	20–29	
Lignites	8–20	
Peats	6–8	
Crude oils	42–44	
Natural gases		29–39

Source: Derived from Smil (1991).

A5.2 Modern Blast Furnaces

Large blast furnaces are undoubtedly among the most remarkable artifacts of industrial civilization. They are slightly cylindrical shafts with a height of just over 30 meters and an internal volume of between 2000 and 4000 cubic meters (m^3). Their main sections, from the bottom up, consist of the circular hearth where the liquid iron and slag collect between tappings, which contains from 15 to more than 40 tuyeres that admit a pressurized, hot, oxygen-enriched air blast (at 3–4 times the atmospheric pressure and over 1200 °C) into the furnace; the bosh, a rather short, truncated, outward-sloping cone containing the furnace's highest temperatures (more than 1600 °C); the slightly narrowing shaft, where the countercurrent movement of downward moving ore, coke, and limestone and upward moving hot carbon monoxide–rich gases reduce oxides into iron; an apparatus for charging raw materials into the furnace; and the pipes for gathering waste gases, which are cleaned and used to preheat the blast or to generate electricity.

These furnaces may produce hot metal continuously for up to ten years before their refractory brick interior and carbon hearth are relined. Molten metal is released regularly through tapholes closed with a clay plug and opened up by a special drill, and the slag (often used by the construction industry) is removed through cinder notches. The mass and energy flows needed to operate large blast furnaces are prodigious (Peacey 1979; Sugawara et al. 1976). A furnace producing 10,000 tonnes of metal a day uses 8 million tonnes of raw materials and an equivalent of nearly 2.5 million tonnes of good steam coal every year.

A5.3 Energy Inputs in Modern Farming

In 1990 the global total of agricultural energy inputs was equivalent to about 300 million tonnes of oil, or less than 5 percent of the total worldwide primary energy consumption. Fertilizers accounted for half of the total, production of field machinery and fuels and spare parts for about 40 percent, and pesticide synthesis and irrigation for about 5 percent each. In relative terms, these inputs prorated to the equivalent of nearly 200 kg of oil for every hectare of land under annual or permanent crops. The difference between the rich and poor countries was relatively small: The former used the equivalent of around 250 kg of oil, whereas the latter used the equivalent of around 180 kg.

Inputs for individual crops and animal products range widely (Pimentel 1980; Fluck 1992a). They were the equivalent of less than 150 kilograms of oil per hectare for Canadian wheat, but ten times that amount for heavily fertilized and irrigated Nebraska corn. Because of their heavy fertilization and irrigation needs, vegetables and fruit crops need far higher energy inputs than grains. Beef production is the most energy-intensive form of animal husbandry. Pork requires much less energy because the animals are ready for slaughter in just six months. Broilers are by far the most efficient converters of feed into lean meat, and also the only species whose efficiency of feed conversion has been substantially raised with breeding. Milk is the least energy-intensive animal food.

In spite of higher crop yields and higher productivities in animal husbandry, rising fuel and electricity inputs have led to falling energy ratios. Traditional agricultures traded food energy inputs in animate labor for harvested crops with ratios ranging mostly between 10 and 30 (see Chapter 3). In contrast, modern cropping systems exchange fossil fuels and electricity for food harvests with ratios well below 10, and in animal husbandry these returns are almost always below 1.

But unlike the ratios in traditional farming, the energy input/output ratios for modern farming are inappropriate and misleading (Fluck 1992b). Inclusion of fuels and electricity in the ratio implies their actual conversion into food outputs. Obviously, this is not the case, as those inputs merely enhance a crop's photosynthetic potential.

Moreover, if society was just concerned about maximizing energy returns, farmers would only cultivate the two highest-yielding crops: corn in temperate regions, and sugar cane in the tropics. Of course, crops are grown not just for their gross energy content but for their unique combinations of nutrients, processing and storage possibilities, and taste.

A5.4   Power Requirements of Common Electrical Appliances and Electronic Devices

Item	Power (watts)
Household appliances	
Stoves	8000–12,000
Hot water heaters	1500–4500
Toasters	600–1200
Microwave ovens	600–1200
Washing machines	300–800
Clothes dryers	300–1000
Refrigerators	200–600
Electronic devices	
TVs	100–300
Compact disc players	20–40
Cassette decks	20–40
Personal computers	20–400

A5.5 Power Densities of Fuel and Food Supply

If the per capita food intake in a traditional city averaged about 9 megajoules per day (MJ/day), this food came overwhelmingly from plant foods, and typical grain yields averaged just 750 kilograms per hectare (kg/ha), a city of 500,000 people would have required about 150,000 hectares of cropland. In a cold climate annual fuel (wood and charcoal) needs would have reached about 2 tonnes per capita. If supplied on a sustainable basis from forests or from fuelwood groves with annual yields of 10 tonnes per hectare (t/ha), around

100,000 hectares would be needed to fuel the city. A densely populated city of that size could have occupied as little as 2500 hectares and thus would have had to rely on an area about 100 times its size for its food and fuel.

In terms of average power densities, this example implies about 25 watts per square meter (W/m^2) for total energy consumption and 0.25 W/m^2 for the supply. The actual range of power densities was fairly large. Depending on their food intakes, cooking practices, heating needs, small manufactures, and combustion efficiencies, total energy consumption of preindustrial towns and cities prorated mostly to between 5 and 30 W/m^2 of their area. The sustainable production of fuel from nearby forests and woodlots yielded anywhere from 0.1 to 1 W/m^2. Consequently, cities commonly had to rely on cropped and wooded areas 50 to 150 times their own size.

Although they use fuels with much higher efficiency, modern cities, with their high concentrations of housing, factories, and transport, consume fuels with power densities ranging from 15 (in sprawling, warm-climate places) to 150 (in cold climates with heavy industries) W/m^2 of their area. However, both the coals and the crude oils supplying these needs are extracted with power densities ranging usually between 1000 and 10,000 W/m^2. This means that a typical industrial city relies on a coalfield or oilfield occupying an area equivalent to no more than 10 percent and as little as one-tenth of 1 percent of its built-up area. As far as the food is concerned, a modern city of 500,000 people consuming 11 MJ per capita daily (with one-third of this amount coming from animal foods requiring, on the average, four times their energy value in feed) could grow enough crops on only about 70,000 hectares—less than half the amount needed for the traditional city—even if the mean yield was just 4 t/ha.

A5.6 Importance of Electricity in Households

The liberating effects of electricity are unforgettably illustrated in an unexpected source: Robert Caro's (1982) first volume of Lyndon Johnson's biography. As Caro perceptively points out, it was not the shortage of energy that made life in Texas Hill County so hard (households had plenty of wood and kerosene), but the absence of electricity. In a moving, almost physically painful account, Caro describes the drudgery, and danger, of ironing with heavy wedges of metal heated on wood stoves, pumping and carrying water for cooking, washing, and animals, grinding animal feed, and sawing. These burdens, falling largely on women, were much harder than the typical labor requirements in poor countries because the Hill County farmers strove to maintain much higher standards of living and run much larger farming operations than peasants in Asia or Latin America. For example, a family of five would need nearly 300 tonnes of water per year, entailing an annual expenditure equivalent to more than 60 eight-hour days and requiring family members to walk about 2500 kilometers. Not surprisingly, nothing could have been so revolutionary in the lives of these people as the extension of transmission lines and the introduction of electric pumps.

A5.7 Destructive Power of Nuclear Weapons

Physical data on the bomb that destroyed Hiroshima at 8:15 A.M. on August 6, 1945, are known in great detail (Committee for the Compilation 1981). The temperature at the point of explosion, 580 meters above the ground, was several million °C, compared to 5000 °C for conventional explosives. The fireball expanded to its maximum size of 250 meters in 1 second, the highest blast velocity at the hypocenter was 440 meters per second (m/s), and the maximum pressure reached 3.5 kilograms per square centimeter (kg/cm^2). The explosion was equivalent to 12,500 tonnes (12.5 kilotonnes [kt]) of TNT. Expressed in standard energy units, the amount that it released totaled 52.5 terajoules (TJ); about half of this amount went into the blast itself and 35 percent went into thermal radiation. These two effects caused a large number of instant deaths. Ionizing radiation caused both instant and delayed casualties. The Nagasaki bomb, at 22 kt of TNT, was slightly more powerful.

These weapons appear minuscule compared to the most powerful thermonuclear bomb tested by the Soviets over the Novaya Zemlya on October 30, 1961: This bomb was equivalent to 58 million tonnes (58 megatonnes [Mt]) of TNT (Sykes and Davis 1987). Less than 15 months later, Nikita Khrushchev revealed that Soviet scientists had built a 100-Mt bomb. Typical warheads on intercontinental missiles are much less powerful, between 100 and 1000 kt, but up to 10 of them can be carried by such missiles as the U.S. submarine-launched Poseidon or the Russian SS-11. Computer models of a full-scale thermonuclear exchange between the United States and the Soviet Union assumed a combined power equivalent to a few thousand Mt. By 1990 the total explosive power of U.S. and Soviet nuclear warheads of all kinds was equivalent to just over 10,000 Mt.

A5.8 Greenhouse Gases

CO_2 is the most abundantly produced greenhouse gas—but also a critical natural constituent of the atmosphere: It is not only the source of carbon for all photosynthesis but also an essential regulator of planetary temperature. In the absence of the gas the Earth's surface would be about 33 °C colder, and hence unable to support a highly diversified biosphere. The gas is essentially transparent to incoming, short-wave solar radiation but a vigorous absorber of the outgoing, long-wave terrestrial radiation, which it then emits both to space and back to the surface. Since the middle of the nineteenth century the natural fluctuations of atmospheric CO_2 have been overwhelmed by the growth of fossil fuel combustion and by the concurrent extensive conversion of grasslands and forests to farmland and built-up areas. CO_2 levels rose from 280 parts per million (ppm) during the eighteenth century to about 350 ppm by 1990 (Boden et al. 1990).

Methane (CH_4) levels roughly doubled during the industrial era, from about 0.8 ppm at the beginning of the nineteenth century to about 1.7 ppm in the late 1980s. Sources of CH_4 include anaerobic fermentation of solid wastes in landfills and garbage pits and of organic matter in flooded farm soils, enteric generation of the gas by livestock, and direct emissions from coal mines, natural gas wells, and pipelines. From 1977 to 1990, concentrations of nitrous oxide (N_2O) rose from about 300 parts per trillion (ppt) to around 310 ppt. The principal sources of N_2O are fertilizers and biomass and fossil fuel combustion. Chlorofluorocarbons (CFCs) have been used since the 1930s as refrigerants, aerosol propellants, foaming agents, and cleansers. Their levels roughly doubled between the late 1970s and 1990 (Thompson et al. 1990).

CH_4, N_2O, and CFCs are much more efficient absorbers of the outgoing long-wave radiation than CO_2 is. One molecule of methane will absorb about 70 times more long-wave radiation than a molecule of CO_2, and the analogical multiples are about 250 for N_2O and on the order of 10,000 for CFCs (World Resources Institute 1990). As a result, in spite of its huge emissions CO_2 accounted for slightly less than 60 percent of the total effect in 1990, CFCs for roughly a quarter, and CH_4 for one-sixth.

6

Energy in World History

Since energy is an essential ingredient in all terrestrial activities, organic and inorganic, it follows that the history of the evolution of human culture must also be a history of man's increasing ability to control and manipulate energy.

—M. King Hubbert, *Energy Resources* (1962)

ALL NATURAL PROCESSES and all human actions involve transformation of energy. Civilization's advances can be seen as a quest for higher energy use converted into increased food harvests and greater mobilization of materials. These accomplishments result in larger populations organized with greater social complexity and enjoying a rising quality of life. Outlining the milestones of this history in terms of dominant prime movers and fuels is fairly straightforward. Nor is it difficult to recount the most important socioeconomic consequences of these technical changes.

What is much more challenging is finding a sensible balance between seeing history through the prism of energy imperatives—and paying proper attention to a multitude of non-energy factors that initiated, controlled, shaped, and transformed human use of energy. Even more fundamentally, it is also necessary to address the basic paradox of energy's role in life's evolution in general and in human history in particular. All living systems are sustained by incessant imports of energy, and this dependence necessarily introduces a number of fundamental constraints. But these life-sustaining energy flows cannot explain either the very existence of organisms or the particular complexities of their organization.

Grand Patterns of Energy Use

The most salient aspect of the ecologic dimension of culture, looked at over millennia of cultural evolution, is the general correspondence between the size and density of culture-bearing populations on the one hand and the amount of potential energy per capita that must be "captured" from the environment and transformed into materials and energy forms on the other.

—Hoyt Alverson, *Culture and Economy* (1986)

Long-term patterns relating human accomplishments to changing prime movers and dominant fuels are perhaps best revealed when viewed in terms of energy eras and transitions. This approach must eschew rigid periodization as well as any categoric statements about the universality of underlying processes. There are too many national and regional particularities driving and shaping such complex changes. In spite of enormous differences in cultural and political settings, there is actually more scope for generalizing about the socioeconomic consequences of these fundamental energetic changes. Identical, or at least very similar, fuels and prime movers have left many repeatedly identifiable imprints on the organization of settlements, production, and warfare, on material accomplishments, and on intangible aspirations.

What recurs again and again is the enormous gap between the traditional societies, governed by animate prime movers and biomass fuels, and the industrial civilization, energized by fossil fuels and electricity. This is true when looking at the level of overall economic output or average individual standards of living, at labor productivities or common travel speeds, and at the capacities to build or the means to destroy. Modern civilization has engineered a veritable explosion of energy use and has extended human control over inanimate energies to previously unthinkable levels. These gains made it fabulously liberating and admirably constructive—but also uncomfortably constraining and horribly destructive.

Energy Eras and Transitions

As a historical tool, the energy transition has much to recommend it. In the broadest sense, the concept can provide a necessary focus for understanding the evolution of human material culture, economic growth and development and, possibly, social organization.

—Martin V. Melosi, *Energy Transitions in the Nineteenth-Century Economy* (1982)

Any realistic periodization of human energy use must take into account both the dominant fuels and the leading prime movers. This need disqualifies the two conceptually appealing divisions of history into just two distinct energy eras. Animate versus inanimate contrasts the traditional societies, where muscles were the dominant prime movers, with modern civilization, dependent on fuel- and electricity-powered machines. But this division misleads both about the past and the present. In a number of old high cultures two classes of inanimate prime movers, waterwheels and windmills, were making a critical difference many centuries before the advent of fossil-fueled machines.

The rise of the West owes a great deal to a powerful combination of two inanimate prime movers: effectively harnessed wind and gunpowder, that is, oceangoing sailing ships equipped with heavy guns (McNeill 1989). And the cleavage between animate and inanimate prime movers has been fully accomplished only among the richest fifth of humanity. Substantial reliance on heavy human and animal labor is still the norm throughout the rural areas of Asia, Africa, and Latin America.

The second simplification—renewable versus nonrenewable—captures the basic dichotomy between the millennia dominated by animate prime movers and biomass fuels and the more recent past heavily dependent on fossil fuels. Actual developments have been more complex. The biomass supply in wooden-era societies was not a matter of assured renewability. Excessive tree cutting followed by destructive soil erosion on vulnerable slopelands destroyed the conditions for sustainable forest growth over large areas of the Old World. And in today's fossil fuel–dominated world, water power generates one-fifth of all electricity and millions of farmers in poor countries still rely on human and animal labor for field work and for the maintenance of irrigation systems.

Clear divisions into specific energy eras are unrealistic not only because of obvious national and regional differences in the time of innovation and widespread adoption of new fuels and prime movers but also because of the evolutionary nature of energy transitions (Melosi 1982). Established prime movers or fuels can be surprisingly persistent. New energy sources or techniques become dominant only after long periods of gradual diffusion. The combination of accessibility and cost explains most of this inertia. As long as the established sources are readily available and profitable, their substitutes will advance only slowly. This is true even when they come with some clearly superior attributes.

Examples abound. Roman water mills were first used during the first century B.C., but they became really widespread only about 500 years later. Even then their use was almost completely limited to grain grinding. As Moses Finley (1965) noted, freeing slaves and animals from their drudgery was not a powerful enough incentive to inspire a rapid diffusion of water mills. By the end of the sixteenth century, circumnavigation of the Earth by sailing ships became almost commonplace—but the Venetian Republic was still building large trading and war galleons powered by scores of oarsmen. Draft animals, water power, and steam engines coexisted in parts of industrializing Europe for more than a century. In the wood-rich United States, coal surpassed fuelwood combustion, and coke became more important than charcoal, only during the 1880s.

Only suggestive approximations are possible in charting long-term patterns of prime mover deployment in the Old World's preindustrial societies. The most remarkable feature is the long dominance of human labor (see Figure 6.1). Human muscles were the only source of mechanical energy from the beginning of hominid evolution until the domestication of draft animals, which started more than 10,000 years ago. Human labor was enhanced by a growing number of better tools. In contrast, animal work throughout the Old World remained limited for millennia by poor harnessing and inadequate feeding. Draft beasts were alto-

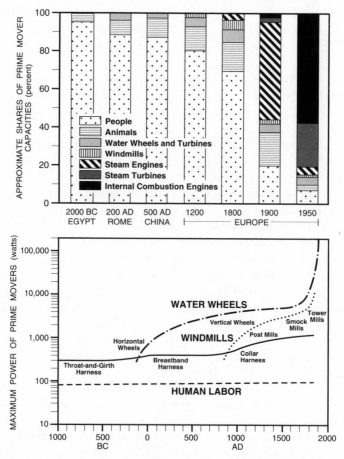

FIGURE 6.1 Prime mover capacities and maximum outputs. The prolonged dominance of human labor, the slow diffusion of water- and wind-driven machines, and the rapid post-1800 adoption of engines and turbines are the three most remarkable features in the history of prime movers (*top*). These realities greatly limited the average unit power of all common pre-nineteenth-century prime movers (*bottom*). Both graphs give approximations based on a wide variety of historical and statistical sources.

gether absent in the Americas and Oceania. Human muscles thus remained indispensable prime movers in preindustrial societies.

A remarkable dichotomy characterized the use of human labor in all ancient civilizations. In contrast to its massed deployment in construction, old high cultures—no matter if based on slave, corvée, or mostly free labor—never took steps to the really large-scale manufacture of goods. Atomization of production remained the norm (Christ 1984). The Han Chinese mastered some potentially large-scale production methods. Perhaps most notably, they perfected the casting of iron suitable for mass-producing virtually identical multiple pieces of small metal articles from a single pouring (Hua 1983). But the largest Han kiln discovered to date is just 3 meters wide and less than 8 meters long.

Outside Europe and North America, relatively small-scale, artisanal manufactures remained the norm until the twentieth century. The lack of inexpensive land transportation was obviously a major factor militating against mass production. Costs of distribution beyond a relatively small radius would have surpassed any economies of scale gained by centralized manufacturing.

And many ancient construction projects did not really require extraordinarily massive labor inputs either. Several hundred to a few thousand corvée laborers working for only two to five months every year could erect enormous religious structures or defensive walls, dig long irrigation and transportation canals, and build extensive dykes over a period of just twenty to fifty years. But we know that many stupendous projects were under construction for much longer periods. Ceylon's Kalawewa irrigation system took about 1400 years to build (Leach 1959). Piecemeal construction and repairs of China's Great Wall extended over an even longer period of time (Waldron 1990).

The first inanimate prime movers—waterwheels and windmills—began to make a notable difference in parts of Europe and Asia only during the second half of the first millennium A.D. Gradual improvements of these devices replaced and speeded up many tiresome, repetitive tasks, but the substitution for animate labor was slow and uneven. In any case, except for pumping water, waterwheels and windmills could do little to ease field tasks. Even during the steam engine era animate labor remained indispensable for extracting and distributing fossil fuels and in performing countless manufacturing tasks. In farming it dominated field work throughout the nineteenth century. In the Americas it also converted most of the currently cultivated land from forests, grasslands, and wetlands.

In the West animate labor was finally displaced by electric motors in manufacturing and by internal combustion engines (tractors, combines) in agriculture. There were no inanimate prime movers in the ancient Middle East or in the pre-Columbian Americas. The contribution of windmills and watermills in dynastic China remained marginal in comparison with human labor, and the aggregate power of the latter also greatly surpassed that of draft animals. Similar conditions prevailed in India and in classical Europe.

The dominance of human muscles limited the most commonly deployed single labor units to between 60 and 100 watts. This means that, in all but a few exceptional circumstances, the highest power concentrations of human labor under a single command (hundreds to thousands of laborers at construction sites) reached no more than 10,000 to 100,000 watts in sustained effort. An ancient or medieval master architect or canal builder thus controlled energy flows at best equivalent to that of a single engine powering today's small earth-moving machines.

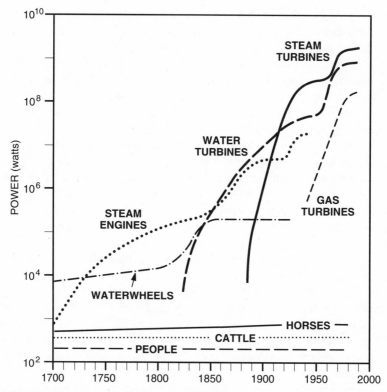

FIGURE 6.2 Maximum capacities of common prime movers. This graph shows that the largest electricity-generating steam turbines are more than a million times more powerful than the strongest draft animals. Steam engine ratings surpassed waterwheel ratings before the middle of the eighteenth century. A hundred years later, water turbines became the single largest prime movers. Steam turbines have remained unsurpassed since the second decade of the twentieth century.

Even small and poorly harnessed draft animals were roughly three times as powerful as an average working man. But long before the maximum power of working animals was itself tripled (by strong horses with collar harness), waterwheels became the most powerful single-unit prime movers. The subsequent development of waterwheels was slow. The first tenfold increase of highest capacities took about 1000 years, the second one about 800. Their peak unit power was surpassed only by steam engines in the late eighteenth century.

Transitions to more powerful prime movers of the industrial era can be traced quite accurately in terms of both typical and maximum capacities (see Figure 6.2). Steam engines were by far the most powerful units until the middle of the nineteenth century. Water turbines gained a short-lived primacy between 1850 and 1910. Since that time steam turbines have been the most powerful single-unit prime movers. The power envelope connecting the peak prime mover capacities thus moved from roughly 100 watts of sustained human labor to about 300 watts for draft animals sometime during the third millennium B.C. The line rose to about 5000 watts in horizontal waterwheels by the end of the first millennium A.D. By 1800 it surpassed 100,000 watts in steam engines, and it was 100 times higher in water turbines a century later. Finally, it has reached a plateau at more than 1 billion watts in the largest steam turbines after 1960.

A different perspective is gained by looking at total prime mover capacities. After 1700 the basic global pattern can be reasonably approximated, and accurate historic statistics make the retrospective easy for the United States (see Figure 6.3). By 1850 animate labor still accounted for more than four-fifths of the world's prime mover capacity. Half a century later its share was about three-fifths, with steam engines supplying about one-third. By 1990 all but a small fraction of the world's available power was installed in internal combustion engines and electricity generators. U.S. prime mover substitutions predated these global changes.

Of course, internal combustion engines—whether in vehicles, tractors, combines, or pumps—are rarely deployed in such a sustained manner as electric generators. Both automobiles and farm machines usually operate less than 500 hours a year, compared to more than 5000 hours for turbogenerators. Consequently, in terms of actual energy production, the global ratio between internal combustion engines and electricity generators was about 2:1 in 1990.

Meanwhile, both the unit power of inanimate prime movers and their total capacity grew. Their mass to power ratios have declined while their conversion efficiencies have increased. The first trend has brought progressively lighter and hence more versatile fuel convertors (see Figure 6.4). The earliest steam engines, although more powerful than horses, were exceedingly heavy. More than two centuries of subsequent development lowered their mass to power ratio to about one-tenth of initial values. The mass to power ratio of internal combustion engines declined by two orders of magnitude in less than fifty years. This precipitous drop opened up the way for affordable mechanization of road transport and made aviation possible. Gas turbines carried these improvements by almost another two orders of magnitude, making speedy and large-scale air travel possible.

The efficiencies of prime movers are limited by fundamental thermodynamic considerations, but technical advances have been narrowing the gaps between best performances and theoretical maxima. The average efficiencies of steam-driven machines rose from a fraction of a percent for Thomas Savery's primitive engines to just over 40 percent for the large turbogenerators of the 1990s (see Figures 5.3 and 5.7). Only marginal improvements are now possible for turbogenerators, whether steam- or water-driven. Similarly, the best combustors perform close to theoretical limits. Both large power plant boilers and household natural gas furnaces may be up to 95 percent efficient. In contrast, the everyday performances of internal combustion engines, the prime movers with the largest

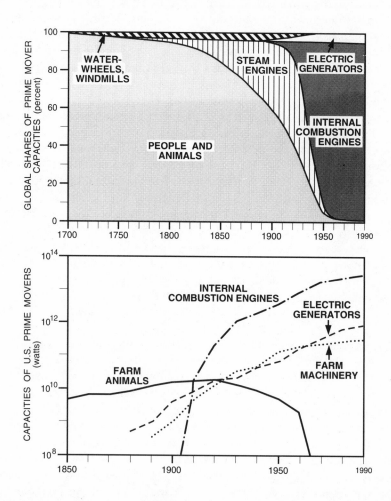

FIGURE 6.3 Global prime mover shares, 1700–1990, and capacities of U.S. prime movers, 1850–1990. Global prime mover shares in 1700 were only marginally different from those of 500 or even 1000 years before. In contrast, by 1950 all but a very small fraction of the world's available power was installed in internal combustion engines and steam and hydro turbines (*top*). Disaggregated U.S. statistics (*bottom*) show this rapid transformation with greater detail and accuracy. *Sources:* Top graph estimated and plotted largely from data in UNO (1956; 1976; 1992) and Mitchell (1975); bottom graph plotted from data in U.S. Bureau of the Census (1975) and various issues of *The Statistical Abstract of the United States.*

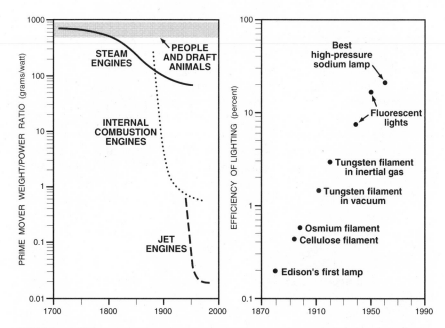

FIGURE 6.4 Prime mover weight/power ratios, 1700–1990, and improvements in lighting, 1870–1990. Every new energy convertor has eventually become much lighter and more efficient. The steady declines in the weight/power ratios of leading prime movers (*left*) means that the best internal combustion engines, now by far the most common prime movers worldwide, deliver the same power with less than 1/1000 of the weight compared to draft animals or early steam engines. Improvements in lighting efficiency (*right*) have been no less remarkable. The best lamps today convert about 100 times more electricity into light than Edison's first light bulb did about a century ago.

aggregate installed power, are still very low. Poorly maintained car engines often perform at less than one-third of their theoretical maxima.

Improvements in lighting efficiency have been equally impressive. Candles converted a mere 0.01 percent of the burning paraffin into light. Edison's first light bulbs, using oval loops of carbonized paper secured by platinum clamps to platinum wires sealed through the glass, were about twenty times more efficient than candles. Osmium filaments, introduced in 1898, converted nearly 0.6 percent of electric energy into light. This rate was more than doubled after 1905 with tungsten filaments in a vacuum, and then doubled again when inert gas was used to fill bulbs. In 1939 the first fluorescent light pushed the efficiency above 7 percent. By 1990 the best high-pressure sodium lamps converted just over 20 percent of electric energy into light (Weast 1992).

Technical advances improved the efficiencies of countless individual convertors and hence raised the amount of useful energy available per capita. But this clear historical trend does not mean that humanity has been using energy in a progressively more rational manner. I suspect that if we could reconstruct in detail the relative shares of surplus energies devoted by nations to preparation for war and to actual fighting, we would find many similarities among many traditional and modern societies—and certainly no encouraging trend.

Every traditional society also diverted a great deal of its energy surplus into pursuits that did not increase its capacity to feed larger populations or to improve their physical quality of life. But monumental structures, the most spectacular results of all unproductive diversions, cannot be seen as simply wasted repositories of scarce resources. Besides embodying deep religious beliefs, they also helped to form social cohesion by encouraging awe, respect, humility, contemplation, and charity. In comparison, it is hard to avoid the conclusion that the label of wasteful energy diversions applies much more readily to the extravagant, and only rarely beautiful, structures modern societies build to warehouse their monies or to watch assorted professional sports. The waste is even clearer with so many high-energy pastimes of modern societies, ranging from jet flights to remote beaches to speedboating and snowmobiling.

On a more mundane level, the diversity of the made world has always been far in excess of physical needs, providing an admirable testimony to the creativity of the human mind (Basalla 1988). But in countless instances, modern society has carried this differentiation to excessive lengths. Do we need more than 100 recordings of Vivaldi's *Four Seasons* or more than 600 types of passenger cars to choose from? Such excessive diversity results in considerable misallocation of energies.

Again, only rough approximations are possible in presenting long-term patterns of the Old World's primary energy consumption (see Figure 6.5). The dominance of biomass energies, above all fuelwood, came to an end only late in the nineteenth century. At that time charcoal, the indispensable fuel in metallurgy and the preferred one for clean indoor burning, was also rapidly replaced by coke, coal, and gas. In cases where the basic energy statistics are known, it is possible to quantify the transitions and to discern long, and remarkably regular, substitution waves. For global fuel consumption, such statistics only became available during the nineteenth century. Substitution rates have been slow, but, considering the variety of intervening factors, they tend to be surprisingly similar. About a century is needed for a new source of primary fuel to reach half of the global market.

In spite of some important continental and regional differences, typical levels of fuel consumption and prevailing modes of prime mover use in old high cultures were fairly similar. If there is an ancient society to be singled out for its notable advances in fuel use and prime mover development, it must be Han China (207 B.C.–A.D.220). Han dynasty innovations were adopted elsewhere only centuries, or even more than a millennium, after their invention. The most notable contributions of the Han Chinese included the use of coal in iron making, innovations in drilling for natural gas and in making steel from cast iron, the widespread use of curved moldboard iron plows, and the invention of the collar har-

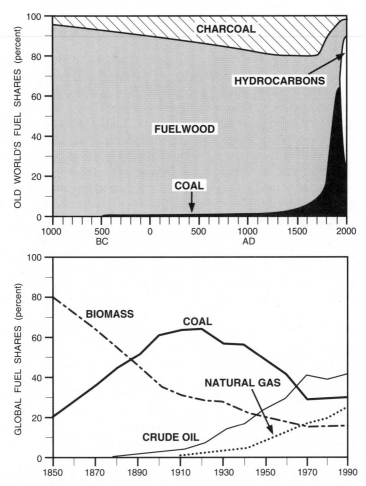

FIGURE 6.5 Fuel shares, 1000 B.C.–1990. Approximate estimates chart the shares contributed by major fuels to the Old World's primary energy supply during the past 3000 years (*top*). The bottom graph reveals fairly regular waves of fuel substitutions worldwide after 1850. *Sources:* UNO (1956; 1976; 1992).

ness and the multitube seed drill. There was no similar cluster of such key advances for more than a millennium!

The early centuries of Islam brought some innovative designs of water-raising machines and windmills, and the realm's maritime trade benefited from an effective use of triangular sails. But the Islamic world did not introduce any radical innovations in fuel use, metallurgy, or animal harnessing. Medieval Europe, however—borrowing eclectically from earlier Chinese, Indian, and Muslim accomplishments—started to innovate in a number of critical ways ranging from using inanimate prime movers to inventing new metallurgical techniques.

Gradually, the combination of crop rotations, better animal feeding, more efficient harnessing, and horseshoing raised the importance of draft horses and improved labor productivity. But what really set European medieval societies apart from their classic ancestors was the widespread reliance on the kinetic energies of water and wind. These flows were harnessed by increasingly more complex machines to provide unprecedented concentrations of power for scores of applications. By the time of the first great Gothic cathedrals the largest waterwheels rated close to 5000 watts, an equivalent of more than three scores of men. Long before the Renaissance some European regions became critically dependent on water and wind, first for milling grain, then for fulling cloth and producing iron. This dependence also helped to sharpen and diffuse many mechanical skills throughout Western civilization.

In late medieval and early modern times Europe was thus a place of broadening innovation, but, as attested by reports of contemporary European travelers admiring the riches of the heavenly empire, the overall technical prowess of contemporary China was certainly more impressive. Those travelers could not know how soon the reverse would be true. By the end of the fifteenth century it was Europe that had embarked on a road of accelerating innovation and expansion—and the elaborate Chinese civilization was about to start its long and deep technical and social involution.

It did not take very long for Western technical superiority to transform European societies and extend their reach to other continents. By 1700 typical levels of energy use, and hence of material affluence, were still broadly similar in China and Western Europe. Then the Western advances gathered speed. In the energy realm they were demonstrated by the combination of improving crop yields, stronger and better harnessed animals, new metallurgical techniques, better navigation, new weapon designs, and a keenness for trade and experimentation. By 1850 the two societies belonged to two different worlds. By 1900 they were separated by an enormous performance gap: Western European energy use was at least four times the Chinese mean.

The foundations of this success go back to the Middle Ages. Christianity's favorable effect on technical advances (notably the recognition of the dignity of manual labor) and medieval monasticism's quest for self-sufficiency were important ingredients of this process (White 1978; Basalla 1988). The period of very rapid advances after 1700 was ushered in by ingenious practical innovators. But its greatest successes during the nineteenth century were driven by close feedbacks between the growth of scientific knowledge and the design and commercialization of new inventions (Rosenberg and Birdzell 1986).

The energy foundations of nineteenth-century advances included the development of steam engines and their widespread adoption as both stationary and mobile prime movers, the use of coke in smelting iron, the large-scale production of

steel, and the introduction of internal combustion engines and electricity generation. The extent and rapidity of these changes came from a novel combination of
these energy innovations with new chemical syntheses and better modes of organizing factory production. Aggressive development of new modes of transportation and telecommunication was also essential, both for boosting production and
promoting national and international trade.

By 1900, the West, now including the new power of the United States, had accumulated technical and organizational innovations that enabled it to command an
unprecedented share of global energy. With only 30 percent of the world's population, the Western nations consumed about 95 percent of the fossil fuels used
worldwide. During the twentieth century the Western world increased its total
energy use nearly fifteen times over. Inevitably, its claim on global energy use declined in relative terms. By 1990, the West had 20 percent of the global population
and consumed about 70 percent of all primary commercial energy. Europe and
North America, joined by Japan, not only remain the dominant consumers of fuels and electricity but have also retained clear technical leadership.

Surges in energy use raised average per capita consumption levels to unprecedented heights (see Figure 6.6). The energy needs of foraging societies were dominated by the provision of food. Ancient high cultures used the slowly rising availability of energy for better shelters and clothes, transportation (energized by food,
feed, and wind), and a variety of manufactures (with charcoal prominent). Early
industrial societies easily doubled traditional per capita energy use. Most of that
increase went into coal-fueled manufactures and transportation.

After 1970 typical per capita consumption rates were two to four times higher
than during the early stages of modernization. Given the higher average conversion efficiencies, the consumption of useful energy was higher still. Energy used
to provide physical necessities has become a steadily smaller part of the rising
consumption. Production of an enormous variety of goods, provision of countless services, and transportation and leisure activities now consume the bulk of
fuels and electricity in all affluent countries.

Consequences

I suggest that as an idea energy is both amoral and ambiguous. It takes on a
quality of utility or of morality, of evil or good intent, only in context.
—Howard M. Jones, *The Age of Energy* (1971)

The adoption and diffusion of new energy sources and new prime movers inevitably had many economic and social consequences. Prehistoric changes brought
about by better tools, mastery of fire, and better hunting strategies were very slow,
unfolding over tens of thousands of years. The subsequent adoption and intensification of permanent farming lasted for millennia. Its most important conse-

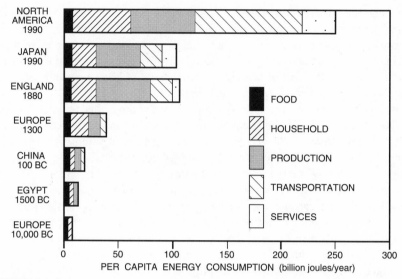

FIGURE 6.6 Comparisons of typical annual per capita consumption of energy during different stages of human evolution. As total consumption has increased, higher shares of energy have been used by households. Shares devoted to food production, manufacturing, service sector, and transportation have shown even higher relative increases. *Sources:* Pre-nineteenth-century values are only approximations; later figures are taken from various national statistical sources.

quence was a large increase in population densities, which led to social stratification, occupational specialization, and incipient urbanization.

The densities of foraging populations spanned a wide range, but—with the exception of some maritime cultures—they never surpassed one person per square kilometer. Even the least productive agricultural method, shifting farming, raised this rate at least ten times. Permanent cropping resulted in yet another tenfold increase. The intensification of traditional farming required higher energy inputs. As long as animate labor remained the sole prime mover of field work, the share of population engaged in crop cultivation and animal husbandry remained very high. The net energy returns of intensive farming, involving irrigation, terracing, multiple cropping, crop rotations, and fertilizations, were generally lower than those in extensive agriculture.

But the most intensive traditional farming methods—most notably Asia's year-round multicropping sustaining largely vegetarian diets—could support relatively high population densities. Averages of more than five people per hectare of cultivated land were common. The resulting occupational specialization, social stratification, and urbanization, as well as world trade, were restricted above all by the limitations of land transportation, especially the slow speeds and low capacities.

In contrast to the slow, cumulative transformations of traditional societies, fossil fuel–based industrialization has brought almost instantaneous socioeconomic changes. The substitution of fossil fuels for biomass, and the later replacement of animate energies by electricity and internal combustion engines, created a new world within just a few generations. The American experience was the extreme example of these compressed changes. More than in any other modern nation, the United States has acquired its power and influence largely through its extraordinarily high use of energy (Schurr and Netschert 1960; Jones 1971). In the 1850s the country was an overwhelmingly rural, wood-fueled society of marginal global import. A century later, after more than tripling its per capita consumption of useful energy and becoming the world's largest producer and consumer of fossil fuels, it was both an economic and military superpower.

The most obvious physical transformations of the new fossil-fueled world have been created by the intertwined processes of industrialization and urbanization. They brought a growing supply and greater variety of food, the mechanization of mass factory production, and better quality of common goods. They introduced new materials (metals, plastics) and greatly intensified trade, transportation, and telecommunication. These developments accelerated every facet of social change. They broke the traditional circle of limited social and economic horizons, improved health, and prolonged lives. They spread both basic literacy and higher education and allowed a modicum of affluence for a rising share of the world's population. They released hundreds of millions of people from hard physical labor and made more space for democracy and human rights.

More productive farming methods and new labor opportunities in expanding industries led to mass migration from villages to cities on all continents. In turn, this rapid urbanization has had an enormous effect on the global use of energy. The infrastructural requirements of urban life increase average per capita energy consumption levels far above rural means even if the cities are not highly industrialized. These relatively high energy density needs could not be supported without cheap means of long-distance transportation of food and fuel, and later without transmission of electricity.

Reliance on electricity rapidly developed into an all-encompassing dependence. Without electricity modern societies could not farm or eat the way they do: It powers compressors in ammonia plants as well as in domestic refrigerators. Modern societies could not prevent disease and take care of the sick, control their transportation networks, or handle their enormous volume of information or urban sewage. And, of course, without electricity they could not operate and manage their industries in ways enabling mass-production of a growing range of high-quality yet affordable goods. This production has erased most of the ancient division between an admirable variety of refined luxury goods produced in small numbers for the richest few and the limited assortment of generally available crude manufactures.

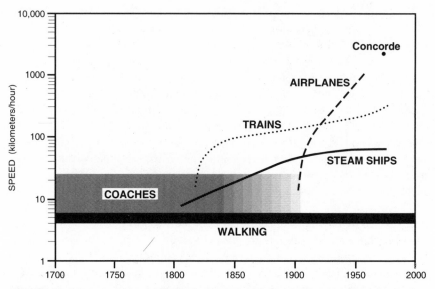

FIGURE 6.7 Maximum mass transportation speeds. Coaches of the prerailway era traveled at maximum speeds of only about 20 kilometers per hour; after just a few decades of locomotive advances, trains could travel at speeds well over 100 km/hour. Large modern jet aircraft routinely fly at speeds close to 1000 km/hour.

A rising share of this advancing output has found its way into the world market. In 1990 foreign trade accounted for about one-fifth of the gross world economic product, compared to less than 5 percent in 1900. This trend has been speeded up by faster and more reliable methods of transportation and by instant electronic telecommunication. Fossil fuels and electricity have moved the world from a mosaic of economic autarkies and limited cultural horizons to an increasingly interdependent whole.

More powerful yet more efficient and lighter mechanical prime movers increased the typical speeds of long-distance modes of land and water transportation more than tenfold and made flying possible (see Figure 6.7). In 1800, horse-drawn coaches usually covered less than 10 kilometers per hour, and heavy freight wagons moved at only half that speed. In 1990 passenger trains and highway traffic could move at speeds far surpassing 100 kilometers per hour, and standard speeds for jet planes (800 to 1000 kilometers per hour) approached the speed of sound.

Increasing speeds have been accompanied by greater ranges and growing capacities in transporting both goods and people. On land, this mechanical evolution has recently reached clear plateaus with multi-axle trucks, unit trains moving bulk materials (carrying up to 10,000 tonnes of coal), and fast electric passenger trains (for up to 1000 people). Supertankers can move up to 500,000 tonnes of crude oil, and the largest planes can carry about 500 people, or nearly

100 tonnes of goods. The greatest standard ranges achievable without refueling now surpass 500 kilometers for a compact car and more than 10,000 kilometers for jumbo jets and large crude oil tankers.

The increased speed and range of transport had its destructive counterpart in the increased speed, range, and effective power of projectiles released by weapons. The killing range of spears was just a few tens of meters, and an expertly fielded spear-thrower increased this distance to more than 100 meters. Good composite bows delivered piercing arrows for distances up to 200 to 300 meters. This was also the range for common crossbows. Various catapults could throw stones of 20 to 150 kilograms some 200 to 500 meters. The reach rose rapidly when muscles were replaced by gunpowder. Just before 1500 the heaviest cannons could fire 140-kilogram iron balls about 1400 meters and the lighter stone balls twice as far (Egg et al. 1971).

By the second decade of the twentieth century, when the ranges of big field pieces reached several tens of kilometers, guns had lost their primacy to bombers. Bomber ranges had surpassed 6000 kilometers with up to 9 tonnes of bombs by the end of World War II (Taylor 1989). In turn, they were surpassed by ballistic missiles. Since the early 1960s these rocket-powered carriers have been able to deliver more and more powerful nuclear bombs with ever greater accuracy either from land-based silos or from submarines to any place on the Earth. From the ancient compound bow to the ballistic missile of the late twentieth century, the increase in range has been about 50,000-fold. Destructive power—calculated in terms of the kinetic energy of arrows and the explosive power of the largest tested nuclear bomb—has risen by sixteen orders of magnitude.

No less profound transformations of the fossil-fuel era have included new structures of social relations. Perhaps the most important one has been a new system of distributing wealth. The change from status to contract in labor relations has led to greater personal and political independence. This transformation brought new work regimens (typically, fixed working hours and multilayered organizational hierarchies) and new social groupings with special interests (labor unions, management, investors). Almost from its beginnings it also introduced new national challenges, above all the need to cope with the extremes of rapid regional industrial growth and chronic economic decline. This disparity continues to plague even the richest nations. New tensions in international relations have been caused by trade barriers, subsidies, and foreign ownership.

New sources of primary energy and new prime movers have also had profound impacts on economic growth and technical innovation cycles. In general, substantial investment is needed to develop new energy sources and new prime movers. In turn, their introduction elicits clusters of gradual improvements and fundamental technical innovations. The classical account of business cycles in industrializing Western countries showed unmistakable correlation between new energy sources and prime movers on one hand and accelerated investment on the other (Schumpeter 1939).

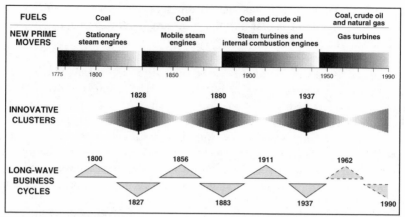

FIGURE 6.8 Comparison of major energy eras, identified by principal fuels and prime movers, with innovation clusters and long-wave business cycles. *Sources:* Innovation clusters according to Mensch (1979); long waves of Western business cycles from Schumpeter's (1939). I have extended both the innovation and business waves to 1990.

The first well-documented upswing (1787–1814) coincided with the spreading extraction of coal and the initial introduction of stationary steam engines. The second expansion wave (1843–1869) was clearly driven by the diffusion of mobile steam engines (railroads and steamships) and advances in iron metallurgy. The third upswing (1898–1924) was decisively influenced by the rise of commercial electricity generation and the rapid replacement of mechanical drive by electric motors in factory production.

The centerpoints of these Schumpeterian upswings are about fifty-five years apart. Fascinating post-1945 research uncovered a great deal of evidence confirming the existence of these cycles in human affairs in general (Marchetti 1986) and in economic life and technical innovation in particular (van Duijn 1983; Freeman 1983; Vasko et al. 1990). These studies indicate that the initial stages of adoption of new primary energies correlate significantly with the starts of major innovation waves. The history of energy innovations also strongly supports a still contentious proposition that economic depressions act as triggers of innovative activity.

The centerpoints of the three temporal innovation clusters identified by Gerhard Mensch (1979) fall almost perfectly into the midpoints of Schumpeterian downswings (see Figure 6.8). The first cluster, peaking in 1828, was clearly associated with the deployment of stationary and mobile steam engines, the substitution of coke for charcoal, and the generation of coal-derived gas. The second one, peaking in 1880, included the revolutionary innovations of electricity generation, electric lighting, telephones, steam turbines, the electrolytic production of aluminum, and internal combustion engines. The third one, clustered around 1937, included gas turbines, jet engines, fluorescent lights, radar, and nuclear energy.

The subsequent extensions of these long cycles also followed patterns of energy use. The post-war economic upswing was associated with the global substitution of hydrocarbons for coal, the worldwide rise of electricity generation (including nuclear fission), mass car ownership, and extensive energy subsidies in agriculture. This expansion was arrested by OPEC's quintupling of oil prices in 1973. The latest wave of innovations has included a variety of highly efficient industrial and household energy convertors and progress in photovoltaics. The rapid diffusion of microchips, advances in computing and optical fibers, and other information techniques will have even greater energy implications.

The economic consequences of the world's prodigious energy use are also reflected by the roster of the world's largest companies (Fortune 1993). In 1992, five of the twenty-five top nonfinancial multinationals were oil companies (Royal Dutch Shell was number one; Exxon was number four). Five were car- and truckmakers (Ford and GM took second and third place), two were electrical equipment suppliers, and three were electronic manufacturers. Concentrated energy supplies have also promoted a concentration of production. Virtually every sector offers good examples.

In 1900, the United States had about 200 car manufacturing companies and France had more than 600 (Byrn 1900). By 1990 there were three American and two French outfits (MVMA 1992). The number of British beer breweries fell from more than 6000 in 1900 to just 142 by 1980 (Mark 1985). But in a number of industries a reverse movement has been under way since the 1970s. This change is largely attributable to a combination of better communication, faster deliveries, and more opportunities to supply specialized demands.

In personal terms, by far the most important consequence of the high-energy era has been the unprecedented degree of affluence and an improved quality of life. Most fundamentally, this achievement is based on an abundant and varied supply of food. People in rich countries have average per capita availabilities much above any realistic needs. Persisting malnutrition, and even hunger, in the midst of this surplus is a matter of distributional inequalities. In physical terms the rising affluence has been manifested most convincingly by drastically lower infant mortalities and longer life expectancies. Intellectually, it has been reflected in higher literacy rates, longer years of schooling, and easier access to a greater variety of information.

Another important ingredient of this affluence has been the use of energy to save time. These uses include much more than a widespread preference for more energy-intensive but faster private cars as opposed to various forms of public transportation. Refrigeration, which obviates daily food shopping, and central heating, which eliminates the job of kindling fires, are also excellent time-saving techniques. In turn, the time gained by these energy investments is increasingly used for leisure trips and pastimes often requiring further considerable energy inputs.

The rise of discretionary, private energy use has been reflected in gradually declining shares of industrial energy consumption and slowly increasing residential and transportation uses. The share of residential consumption is a revealing measure of a country's affluence. Almost certainly it is a more reliable indicator than the gross national product (GNP) per capita. By the late 1980s less than 10 percent of the total energy consumed in China went for residential uses, compared to about 15 percent in Russia, 20 percent in Japan, and about 30 percent in the United States. Regardless of the indicators used, only about one-fifth of the world's more than 150 countries have accomplished the transition to mature, affluent industrial societies supported by a high per capita consumption of energy.

Consequently, the Western miracle of rapid technical innovation resulted in a worrisome global split—an unprecedented level of economic inequality among nations. By 1990 the richest one-tenth of humanity claimed nearly 40 percent of the world's energy. In personal terms this meant that a week's worth of gasoline for a two-car suburban American family was equivalent to the total annual primary energy consumption of a fairly well-off Indian villager!

These disparities have no easy remedies. Even if the requisite resources were readily available, the environmental consequences of raising the rest of the world to the Western level of primary energy consumption would be incalculable. Already, concerns about biospheric integrity have become major considerations in contemplating the future of high-energy civilizations. They range from preservation of biodiversity to the possibility of rapid anthropogenic climatic change.

Between Determinism and Choice

Evolution is neither necessary, in the sense of being predestined, nor is it a matter of chaos or accident.

—Theodosius Dobzhansky
in *Beyond Chance and Necessity* (1975)

Many historical changes are not merely possible outcomes of using particular energies in certain ways—but unavoidable consequences. Different prime movers determine the tempo of life, and different primary energies leave distinct imprints on everyday work and leisure.

A world where people spend their lives breaking clods with heavy hoes, transplanting dripping seedlings, grasping handfuls of stems and cutting them with sickles, husbanding the straw for cooking, and hand-grinding the grain is very different from a world where strong horses pull curved moldboard plows, seeders, and harvesters, where a fuelwood grove yields plenty of wood for large stoves, and where flour is ground by water mills. Laborious harnessing, the clatter of horseshoes, poorly sprung coaches, animal muzzles stuck in feedbags, suburban gardening nourished by horse manure—these images evoke a pace of life profoundly unlike that dominated by the turning of ignition keys, the swishing of radial tires, the smooth and swift ride of sedans, networks of filling stations, and truckloads of lettuce from California or Spain.

It is thus profitable and desirable to view energy use as a principal factor in the analysis of human history. But not as *the* principal factor. The explanatory power of energy as an approach to history must not be exaggerated. Imprecise claims lead to indefensible conclusions. To generalize, across millennia, that higher socioeconomic complexity requires higher and more efficiently used inputs of energy is to describe indisputable reality. To conclude that every refinement of energy flows brought a refinement in cultural mechanisms (Fox 1988) is to ignore a mass of contradictory historical evidence. The only rewarding and revealing way to assess energy's importance in human history is to find a path that neither succumbs to the simplistic, deterministic explanations buttressed by recitals of countless energy imperatives—nor belittles energy use by reducing it to a marginal role compared to other history-shaping factors, be they climatic changes and epidemics or human whims and passions.

Imperatives of Energy Needs and Uses

> *Man is in truth the prisoner of limitations from which he can never escape. These limitations, though they vary in time, are sensibly the same from one end of the earth to the other, and it is they which place their uniform seal on all human experience, whatever the age being considered.*
> —Fernand Braudel, *On History* (1980)

Energy's quintessential role in the physical world and in the maintenance of life must be necessarily reflected by evolutionary and historic developments. The prehistoric development of human societies and the growing complexity of high civilizations have been marked by countless energy imperatives.

The most fundamental physical limit is, of course, incoming solar radiation. This flow keeps the planet's temperatures within the range suitable for carbon-based life and powers atmospheric motion and the water cycle. Temperature, precipitation, and availability of nutrients are the key determinants of plant productivity, but only a part of the newly synthesized biomass is digestible. These realities shaped the basic existential modes, population densities, and social complexities of all foraging societies. In an overwhelming number of cases these groups had to be omnivorous. Most of their food energy had to come from abundant seeds (combining starches with proteins and oils) and tubers (rich in carbohydrates).

Concentrated abundance and the easy accessibility of collectible plants mattered more than their overall biomass and variety. Grasslands and woodlands offered a better existence than dense forests. Killing large (meaty and fatty) mammals brought huge net energy returns, but hunting smaller species was almost always a much less energy-rewarding pursuit than collecting plants. Lipids were the most desirable nutrients commonly in short supply. Their high energy density provided the satisfactory feeling of satiety. These energy imperatives molded collecting and hunting strategies and contributed to the emergence of social complexity.

As long as human, and later animal, muscles remained the only prime movers, all labor rates were determined by metabolic imperatives: by digestion rates of food and feed, by basal metabolic rates and growth requirements of bodies, and by the mechanical efficiency of muscles. Sustained delivery of human power could not be higher than about 100 watts. The efficiency of muscles could not exceed 20 to 25 percent. Only a massing of people or draft animals could overcome these limits, but this solution itself raised problems of effective, coordinated control.

Aggression powered by human muscles had to be discharged either in hand-to-hand combat or by attacks launched stealthily from no farther than a couple hundred meters. For millennia warfare (and hunting) had to be conducted at fairly close quarters. Human anatomy makes it impossible for an archer to exert the maximum force when one extended and one flexed arm are separated by more than about 70 centimeters. This factor limits a draw, and hence the arrow's range. Catapults, tensioned by many hands, boosted the mass of projectiles but did not increase the range of attack.

Although the shift from foraging to farming was driven by a combination of energy-related, nutritional, and social factors, the subsequent intensification of sedentary farming can be explained as a clear energy imperative (Boserup 1965; 1976). When an existing way of food production approaches the physical limits of its performance, the population can either stabilize its size (by controlling births or by outmigration)—or has to adopt a more fruitful system of food production. The onset and duration of successive intensification steps has varied greatly around the world, but in order to harness more of a site's photosynthetic potential every advance demanded higher energy inputs. In return, greater harvests were able to support higher population densities.

Farming intensification also required higher indirect energy investments in breeding and feeding draft animals, in making, bartering, or buying increasingly more complex tools and implements, and in such long-term infrastructural projects as terracing and building irrigation canals, impoundments, granaries, and roads. This intensification led to a growing reliance on sources of energy other than human muscle. Plowing, even on a relatively small scale, is either enormously taxing or outright impossible without draft animals. Milling grains by hand is so labor-intensive that animal, water, and wind power were necessary to process concentrated harvests. Traditional societies also grew to rely largely on animal power and wind for long-distance grain distribution. Moreover, they needed charcoal for manufacturing more durable and more efficient iron tools and implements.

A number of particular energy imperatives have shaped the world of traditional agricultures. Human energy needs set the limits to the use and the feeding of animals (Clark and Haswell 1970). Societies with annual outputs of less than 300 kilograms of unmilled grain per capita could not afford any draft animals unless they had substantial grazing land available. Above that level animals were increasingly used, but initially they were fed almost exclusively with crop residues and by grazing. As the per capita availability of unmilled grain rose above 700 kilograms a year, enough of it was left to feed animals with plenty of concentrates.

Intensity of cultivation, that is, the energy density of food production, also mattered. There was hardly any space for large grazing animals in intensive wet field agriculture. In contrast, the presence of large numbers of cattle, horses, and other domestic animals in land-rich regions influenced both the density of rural populations and the organization of settlements. Plenty of space had to be devoted to barns and stables and to manure storage.

The two agroecosystems embodying these extremes have been the traditional rice-dominated multicropping of China south of the Yangzi and the West European mixed farming with its heavy reliance on animals both for food and draft. Pre-Columbian agricultures were shaped by different energy imperatives. All else being equal, corn yields more than other dryland cereals. This advantage only increases when it is intercropped with beans.

Energy imperatives shaping the nonagricultural activities and structures of traditional societies ranged from basic locational determinants to challenges of effective administration. A larger scale of metal smelting and forging was made possible only by the use of water power. This necessity would have restricted the location of furnaces and forges to mountainous or hilly areas even if a cheap method of transporting ore and charcoal were available. In reality, the limited power of draft animals (and also poor roads) greatly restricted the range of profitable landborne movements of bulky materials. Inefficient charcoaling (with less than a fifth of the initial wood energy converted into smokeless fuel) caused extensive deforestation.

The administration of distant territories, as well as trading and military ventures, were made difficult not only by the slowness but also by the unreliability of both land and sea travel. Sailing from Rome to Egypt, the Roman empire's largest producer of surplus grain, could take as little as a week—and as long as three months or more (Duncan-Jones 1990). And as late as 1800, English ships had to wait sometimes up to three months for the right wind to take them into Plymouth Sound (Chatterton 1926).

Energy imperatives had a profound influence on national and regional fortunes during modern energy transitions. Countries and locales with a relatively easy access to fuels that could be produced and distributed with less energy than the previously dominant source enjoyed faster economic growth with its welcome correlates of greater prosperity and higher quality of life. Conversely, the old source areas were plunged into often chronic economic decline. The earliest nationwide example of this early advantage is the heavy Dutch reliance on peat, which ushered in the republic's Golden Age during the seventeenth century (DeZeeuw 1978).

Only a few generations later an even more far-reaching advantage of this kind was demonstrated with the virtually complete replacement of English wood and charcoal by good-quality bituminous coal and coke. After 1870 that experience was greatly surpassed by the rise of the U.S. economy fueled initially by excellent coals and later by hydrocarbons. Undoubtedly, the successive positions of economic leadership and the international influence of the Dutch Republic, Great Britain, and the United States have been closely related to their early exploitation of fuels extractable with smaller amounts of investment per unit of useful energy.

The dominance of fossil fuels and electricity has created an unprecedented degree of technical, and by a gradual extension, also economical and social uniformity. Even a basic list of universal infrastructures of high-energy civilization is a lengthy one: coal mines, oil and gas fields, thermal power stations, dams, blast furnaces, aluminum smelters, fertilizer plants, countless manufacturing enterprises, railroads, multilane highways, airports, skyscrapers, suburbia. The design, construction, and management of these infrastructures have increasingly originated with a relatively small number of companies supplying the global market with critical machines, processes, and know-how.

The two most obviously negative imperatives of depending on high energy flows is the restriction of choice and degradation of the environment. The first phenomenon is well illustrated by the impossibility of surrendering the high energy subsidies in modern farming without profoundly transforming the whole of modern society. Matching the existing power of American tractors with horses would require building up the draft stock to at least ten times its record numbers of the 1910s. Some 300 million hectares, or *twice* the *total* area of U.S. arable land, would be needed to feed the animals! Naturally, a requisite number of urbanites would have to leave the cities to take care of those animals and fields. And it is not only the rich Western world that cannot go back to traditional farming. China's dependence on fossil energies in farming is even higher than the West's.

Restriction of choice is a paradoxical result of a world dominated by *la technique* (Ellul 1964). This world gives us unprecedented benefits and almost magical freedoms but in return we have to submit to its rules and strictures. Every person now depends on these techniques, but no single individual understands them in their totality: We just follow their dictates in everyday life. Consequences go beyond ignorant obedience; the spreading power of techniques has already made a large part of humanity irrelevant to the production process.

The next logical step is to see this reality as a part of the process leading to the eventual displacement of carbon-based life by machines (Wesley 1974). The evolutionary parallels between living organisms and machines are intriguing. Machines are thermodynamically "alive" and their diffusion conforms to natural selection: Failures do not reproduce, new species proliferate, and they tend toward maximum supportable mass; successive generations are also progressively more efficient (recall all those impressively lower mass to power ratios!), more mobile, and have longer life spans.

These parallels can be dismissed as merely intriguing biomorphizations, but the ascendance of machines has been an undeniable fact. We have already replaced enormous areas of natural ecosystems with the infrastructures needed for making, moving, and storing them (mines, railroads, roads, airports, factories, parking lots); we have spent increasingly greater amounts of time serving them; their waste products have caused extensive degradation of soils, waters, and the atmosphere; and the global mass of automobiles alone is already much higher than that of the whole of humanity. The fact that fossil fuel resources are finite may do little to stop the ascent of machines: In the near term they can adapt by becoming more efficient; in the long term they can rely on renewable flows.

In any case, only a fundamental misinterpretation of clear geological evidence can see in the rising use of fossil fuels a cause for concern about their early exhaustion. Fossil fuel reserves are that small part of the resource base whose extraction is clearly profitable: Their spatial distribution and recovery costs (at current prices and with existing techniques) are known in enough detail to justify their commercial exploitation. As we recover higher shares of originally available resources, the best measure of their availability is the cost of producing additional or marginal units of a mineral. This approach takes into account improvements in exploitation techniques and society's ability to pay the price of recovery (Tilton and Skinner 1987). Exhaustion is not thus a matter of actual physical depletion but rather a burden of eventually insupportable cost increases. There are no sudden ends, only prolonged declines, slow exits, and gradual shifts onto new supply planes.

This understanding is critical in appraising the rise and prospects of fossil-fueled civilization. The existence of finite fossil fuel resources does not imply any fixed dates for the physical exhaustion of coals or hydrocarbons. Nor does this mean an early onset of unbearably rising real costs of recovering these resources and hence the necessity of a rapid transition to a post–fossil fuel era. Reserve estimates and resource appraisals are not certain enough for speculating about the future of fossil fuels. Global demand (highly modifiable) and efficiency of use (greatly improvable) are no less important. Consequently, it is not an early exhaustion of fossil fuels but rather the impact on the habitability of the biosphere that is the most worrisome near- and long-term concern resulting from the world's dependence on coals and hydrocarbons.

Importance of Controls

In dealing with energy, then, we must be interested in triggers as well as in the substantive work done. Triggers are inherently inhibiting mechanisms. They

Adoption of new prime movers and fuels could never have had such far-reaching
consequences without the new inventions that offered societies new modes of
controlling energy conversions. These controls or triggers can open the previously
shut gates and release new flows of energy—or they can increase the overall working rates of established processes, or make them more reliable or more efficient.
They can be simple mechanical devices as well as sophisticated arrangements that
require themselves considerable energy inputs. Or they can be just better sets of
managerial procedures, newly identified and developed markets, or fundamental
political or economic decisions.

No matter how capable, a horse becomes an effectively controlled prime mover
only when a bit is inserted in its mouth and connected to reins held in a rider's
hand; it can pull a battle chariot only with a good light harness; it can be used in
armored combat only when saddled and fitted with stirrups; it can provide heavy
draft only when shod and harnessed with a comfortable collar; and it can be a
part of an efficient team only when the unequal pulls of differently sized animals
are equalized by swingletrees.

The absence of proper controls could mar the performance of otherwise admirable prime movers. The remedy was sometimes very slow in coming. Perhaps the
best illustration of such a failure was the inability to determine longitude. By the
beginning of the eighteenth century, high-masted fully rigged sailing ships were
efficient convertors of wind energy and potent empire-building tools of European
nations. But once on the high seas, their captains still could not find their longitude. Consequently, as summed up in a petition presented by English captains
and merchants to Parliament in 1714, "Many ships have been retarded in their
voyages and many lost."

Because the Earth's rotation amounts to about 460 meters a second on the
equator, finding the longitude required chronometers that would lose no more
than a fraction of a second a week if a ship was to be positioned with an error
smaller than a couple of kilometers after a journey of two to three months. In 1714
an act of the British Parliament offered up to £20,000 for such an achievement.
The reward was finally given to John Harrison in 1773 (Quill 1966).

As far as fuels are concerned, history would have taken a different course if traditional societies had used coal merely as a substitute for wood in open fireplaces,
or if the nineteenth-century quest for crude oil had ended with the production of
kerosene for lighting. In most cases it has not been the access to abundant energy
resources or to particular prime movers that made the long-term difference. Decisive factors were rather the quest for innovation and the commitment to deploying and perfecting new techniques. These factors have determined the energy
efficiencies of whole economies and of particular processes and have made new

conversion techniques safer and more acceptable. Examples of these sometimes striking but often subtle contributions can be found in all energy eras and for all fuels and prime movers.

In a narrow, technical sense, perhaps the most important class of controls includes a variety of feedback devices (Mayr 1970). These devices transfer the information about what is going on in a particular process back to the controlling mechanism, which can then adjust the operation. Early modern Europe gained a decisive leadership in the development of these feedbacks. Among the earliest widely used devices were thermostats (invented in around 1620 by a Dutch engineer, Cornelis Drebbel); fantails (for automatically adjusting windmills, patented in 1745 by an English blacksmith, Edmund Lee); floats in domestic cisterns and in steam boilers (1746–1758); and centrifugal governors (for regulating the power of steam engines, invented in 1789 by James Watt).

An indispensable class of controls embraces the instructions that make it possible to duplicate production and management processes and to turn out standardized goods and services. The rapid development of printing in early modern Europe made an immense contribution in this regard. By 1500 more than 40,000 different books or editions had been published in Western Europe in various quantities for a total of more than 15 million copies (Johnson 1973). The introduction of detailed copper engraving during the sixteenth century, and contemporaneous development of various map projections, were other notable early advances. Another outstanding innovation in this class was a punched card device invented by Joseph Marie Jacquard in 1801 to control loom operations. Before 1900 punched cards were being employed on Herman Hollerith's machines used to process census data. After 1940 punched input controlled first electromechanical and then electronic computers.

Until the end of the nineteenth century, the new controls remained largely mechanical. During the twentieth century, advances driven by a combination of electricity-based techniques and mathematical theories created a vast new field of increasingly sophisticated automatic control. Critical innovations ranged from widespread deployment of radar (in fire control, bombing, missile guidance, and autopilots) to myriads of microchip-based controls in computing, consumer electronics, and industrial processes.

In a wider sense, certainly the most basic control consideration is what societies do with their energy sources and prime movers. How do they apportion them among productive uses and personal consumption? What balance, if any, do they want to strike among the contradictory pulls of autarchy and extensive trade? How open do they want to be to both goods and ideas? How much are they willing to invest in their military forces? How much central control do they wish to exercise? In all of these respects, cultural, religious, ideological, and political constraints, impulses, and leanings have been decisive. Again, numerous examples can be selected from all energy eras. Two momentous contrasts are especially revealing: the first one between Western and Chinese nautical voyages of discovery, the other between the Russian and Japanese approaches to economic modernization.

Transoceanic voyages required not only sails capable of taking ships closer into the wind but also stronger hulls, good sternpost rudders, and reliable navigation devices. The Chinese originated most of these advances and combined them in their great Ming fleets. Just over a century after Marco Polo's return from China these fleets penetrated farther westward from China than any Europeans did eastward at that time. Between 1405 and 1433 the Chinese repeatedly visited the coast of East Africa, and they may have made a foray into the Atlantic (Needham et al. 1971).

Then a sudden involution of the centrally governed empire made further sailings impossible. In contrast, late medieval European ships started out as decidedly inferior prime movers, but the enterprise was sustained by a combination of inquisitive, aggressive, and zealous attitudes. This amalgam of economic, religious, and political aspirations resulted in eventual European domination of the seas and extensive European empires.

The post-1945 economic fortunes of the Soviet Union and Japan demonstrated sharp contrasts: quantity versus quality, autarchy versus commerce, and the role of state as the sole arbiter versus the state as leading catalyst of modernization. Japan had to base its economy on imported fuels. Minor coal resources, limited hydro capacities, and the virtual absence of hydrocarbons left it with no other choice. In order to reduce its vulnerability to import interruptions, it became one of the world's most efficient users of energy. Its governing bureaucracies have promoted state cooperation with industries, technical innovation, and exports. In contrast, the former Soviet Union's rich mineral patrimony made the country the world's largest producer of fossil fuels. But generations of rigid central planning, autarchic Stalinist five-year plans, which continued long after the dictator's death, and excessive militarization of the economy made the country the least efficient user of energy in the industrialized world.

Decisive controls of critical energy flows have often been beyond human influence or have been usurped, or at least heavily affected, by minorities. William H. McNeill (1980) conceptualized these notions in his dual treatment of micro- and macroparasitism. Microparasites—bacteria, fungi, insects—forestall human efforts to secure sufficient food energy. They damage or destroy crops and domestic animals, or they prevent efficient use of digested nutrients by directly invading human bodies.

Macroparasitism has taken a variety of social controls of energy flows relying both on coercion—ranging from slavery and corvée labor to military conquest—and on complex (and partly voluntary) relationships among unequal groups of people. Special-interest groups have become certainly the most important macroparasites in affluent countries. They range from various professional associations and labor unions with restricted entries to oligopolistic industrial cartels and lobbies. By shaping, and often vetoing or disabling, government policies and by fixing prices, these groups work against the optimized use of all resources. Inevitably, they have a notable effect on the development of energy resources and on the efficiency of their use.

Mancur Olson (1982) aptly called such groupings distributional coalitions and pointed out that stable societies will gradually acquire a greater number of such alliances. This argument helps to explain British and U.S. industrial decline as well as the post–World War II successes of Germany and Japan. The organizations that the victorious powers created in the two defeated nations after 1945 were much more inclusive than the prewar arrangements. Energy performance supports this view: Japanese and German economies are clearly less energy-intensive than the British and American ones. This is true not only in aggregate terms but also in most major sectors.

In contrast, the actions of some special groups have opened new energy gates and improved the efficiency of some conversions. Nineteenth-century British emigrants possessed many skills, and their influx into the United States was obviously an important trigger of higher energy flows (Adams 1982). Multinational companies provide another example. Forced to compete globally, they strive to lower the energy intensity of their production, diffuse new techniques, and foster higher energy conversion efficiencies worldwide.

Limits of Energy Explanations

> *Is it possible that the energy flux parameter has now increased to a point that could drive the nonlinear, dynamical process called civilization to chaos? Or is man's nervous system sufficiently advanced to predict future events and establish effective control mechanisms?*
>
> —Ronald E. Fox, *Energy and the Evolution of Life* (1988)

Fernand Braudel (1980) asked if one could really presume to explain civilizations in terms of their environment—and replied in a disdainful negative: "Nothing so crassly material could determine them" (p. 191). Analogically, it is clear that the kinds of prime movers and levels of energy use *do not determine* the aspirations and achievements of human societies. To begin with, there are indisputable natural reasons for this reality. Of course, energy conversions are absolutely essential for the survival of all organisms—but their modification and differential utilization is governed by properties intrinsic to the organisms.

Energy flows do not determine biospheric organization and they cannot explain the existence of organisms and their variability. Or, as Daniel Brooks and E. O. Wiley (1986) put it, living organisms are "Physical systems with genetically and epigenetically determined individual characteristics, which utilize energy that is flowing through the environment in a relatively stochastic manner" (p. 38). In complex human societies energy use is clearly much more a matter of desires than of mere physical needs. The amount of energy at a society's disposal puts clear limits on the overall scope of action, but it may tell us little about the group's basic economic accomplishments or its ethos. Dominant fuels and prime movers are among the most important factors shaping a society—but they do not determine the particulars of its successes or failures.

This limitation is especially clear when one examines the energy-civilization equation. This concept, so pervasive in modern society, equates high energy use with a high level of civilization. The genesis of the link is not surprising. Only the rising consumption of fossil energies has been able to satisfy so many material desires on such a large scale. Greater possessions and comforts have become equated with civilizational advances. This biased approach excludes the whole universe of creative—moral, intellectual, and esthetic—achievements that have no obvious connection with any particular levels or modes of energy use.

Nicholas Georgescu-Roegen (1980) suggested a fine analogy that captures the challenge of historical explanation: Geometry constrains the size of the diagonals in a square—not its color, and "how a square happens to be 'green', for instance, is a different and almost impossible question" (p. 264). And so every society's field of physical action and achievement is bound by the imperatives arising from the reliance on particular fuels and prime movers—but even small fields can offer brilliant tapestries whose creation is not easy to explain. Mustering historical proofs for this conclusion is easy, in matters both grand and small.

Ancient thinkers and moralists of the Middle East, India, and China formulated universal and lasting ethical precepts in low-energy societies often preoccupied with basic physical survival. Classical Greeks often referred to their slaves in terms clearly placing them on the level of working animals (*andrapoda,* man-footed, versus *tetrapoda,* four-footed)—but they gave us the fundamental ideas of individual freedom and democracy. The concurrent advance of freedom and slavery is one of the most remarkable aspects of Greek history (Finley 1959).

Timeless literature, painting, sculpture, architecture, and music shows no correlation with advances in energy consumption. The simplicity of Neolithic cave paintings, the proportions of classical temples, and the haunting sounds of medieval chants are no less pleasing and captivating, no less modern, than the colorful compositions of Joan Miro, the sweeping lines of Kenzo Tange's buildings, or the anguish of Stravinsky's music.

Nor do the great political ideas guiding national histories show any obvious links to energy use. The United States adopted its visionary constitution when the society was a relatively low consumer of wood. Late nineteenth-century Germany embraced aggressive militarism—and, later, fascism—just as it was becoming Europe's leading consumer of coal. The Communist dictatorships of the former Soviet empire used much more energy than Western European democracies—yet they could offer their people much less comfortable lives.

The satisfaction of basic human needs obviously requires certain minima of energy, but international comparisons clearly show a limited capacity to improve those needs with rising energy consumption (Goldstein 1985). Societies focusing more on human welfare than on frivolous consumption can achieve low infant mortalities and high life expectancies while consuming a fraction of the fuels and electricity used by more wasteful nations. Contrasts between Japan and Russia, Costa Rica and Mexico, or Israel and Saudi Arabia make this obvious. In all of these cases the external realities of energy flows have been obviously of secondary importance to internal human motivations and decisions. Identical sources and prime movers have produced fundamentally different outcomes—while highly disparate consumption rates have resulted in surprisingly similar levels of physical quality of life.

The spirit behind the facade of physical reality is equally discernible in the virtual worldwide identity of high-energy structures and processes. Regardless of the universal imperatives of their energy and material inputs and operating requirements, the blast furnaces of Chicago, Ruhr, Donets, Hebei, Kyushu, and Bihar are *not* the same. Their distinctiveness relates to the amalgam of cultural, political, and social settings in which they operate, as well as to the eventual destination and quality of products made from the metal they smelt.

Energy availability is also of limited usefulness in explaining population growth. Most reliable long-term demographic reconstructions for Europe and China show very long periods of slow growth composed of successive waves of expansions and crises caused by epidemics and wars (McEvedy and Jones 1978). Europe's total population during the first half of the eighteenth century was about three times that at the beginning of the common era—but by 1900 it had more than tripled. Improved nutrition had to be a major reason for this rise—but not the only underlying factor (McKeown 1976), as careful reconstructions of average food energy intakes have shown (Livi-Bacci 1991).

Nor is it very helpful to see the rising use of fossil fuels, reflected in better housing, hygiene, and health care, as the key factor driving population growth. Undoubtedly, these changes were of major importance in Europe, but hardly so in contemporary China. In 1700 China's population total was only about three times the Han peak in A.D. 145—but by 1900 it nearly matched European growth by tripling to about 475 million people. Yet during that period there were no major shifts to new energy sources or prime movers and hardly any improvement in average per capita use of biomass fuels and coal.

Not surprisingly, energy considerations are also of limited help in trying to explain some of the greatest recurrent puzzles of history, the collapse of complex societies. Voluminous research on this fascinating and challenging issue (Yoffee and Cowgill 1988; Tainter 1989) has yielded few clearcut answers. The most notable energy-related explanations have cited the effects of widespread ecosystemic degradation brought about by unsustainable farming methods and excessive deforestation. Other common explanations have focused on the difficulty of effectively integrating large empires when only poor methods of land transportation are available and the burden of protecting distant territories (the imperial overstretch syndrome).

But, as attested by scores of different reasons offered to explain the fall of the Roman empire—by far the most studied collapse in history (Rollins 1983)—explanations favoring social dysfunction, internal conflict, invasions, epidemics, or climatic change are much more common. An indisputable fact is that many instances of sociopolitical collapse came about without any persuasive evidence of weakened energy bases. Neither the slow disintegration of the Western Roman empire nor Teotihuacan's sudden demise can be persuasively tied to badly degraded food production capacity, to notable shifts in dominant prime movers, or to dramatic changes in the use of biomass fuels. Conversely, a number of historically far-reaching consolidations and expansions—including the gradual rise of Egypt's Old Kingdom or the Roman republic or the swift spread of Islam or Mongolian rule—cannot be tied to any major changes in prime movers or fuels.

Extreme futures are easy to outline. On one hand, it is conceivable that the understanding now mastered by Western civilization may take it beyond others in its basic behavior patterns. Diffusion of this knowledge can create a true world civilization that, "unlike all that have preceded it, might be perpetually reconstituted and maintained at a level undreamed of in past history" (Melko 1971, 39). On the other hand, it is equally conceivable—even when leaving the possibility of a full-scale nuclear war aside—that the global, high-energy civilization may collapse long before approaching the limits of its resources.

Its gradual dissipation, brought about through degradation of the biosphere beyond sustainable habitability, would now seem to have a higher probability than a sudden demise. This does not mean that a concatenation of social breakdown, violent conflicts, and new pandemics could also not cause a rapid, Teotihuacan-like demise. I will offer no predictions regarding the chances of destructive social dysfunction, worldwide wars, or epidemics. I will merely note the coexistence of two contradictory trends in energy forecasting: chronic conservatism and repeated exaggeration.

The list of failed technical predictions is very long. Some of the choicest items have concerned the development and use of energy conversions (Gamarra 1969). Expert opinion of the day dismissed the possibility of gas lighting, steamships, incandescent light bulbs, gasoline engines, powered flight, alternating current, radio, rocket propulsion, and nuclear energy. This conservatism often persevered even after successful innovations appeared. Transatlantic steamship voyages did not seem possible because it was thought that the vessels could not carry enough fuel for such long trips. In 1904, Octave Chanute, a chronicler of aviation advances, raised the same argument when he dismissed the practicality of long flights carrying passengers or arms. At the time when scores of automakers were turning out more efficient and more reliable cars, Edward Byrn (1900) thought, "It is not probable that man will ever be able to get along without the horse" (p. 271).

The persistence of new energy myths is at least as remarkable as this skepticism. New energies are initially seen to carry few if any problems. They promise abundant and cheap supply opening up a possibility of a near-utopian social change (Basalla 1982). After millennia of reliance on biomass, many nineteenth-century writers saw coal as an ideal energy source and the steam engine as a nearly miraculous prime mover. Heavy air pollution, land destruction, health hazards, mining accidents, and the necessity to turn to progressively poorer or deeper coal reserves soon swept that myth away.

Electricity was the next carrier of unbounded possibilities. Its promise inspired innovators (Edison, Westinghouse, Steinmetz, Ford) and politicians (Lenin, Roosevelt) alike. Its clean form, the "white coal" of hydroelectricity, held a special appeal until the 1950s when its mystique yielded to the unprecedented promise of nuclear power. This energy was eventually supposed to be too cheap to meter, absolutely nonpolluting, and safe. In 1971, Glen Seaborg, chairman of the U.S. Atomic Energy Commission, forecast that half of America's electricity generating capacity would be nuclear by the year 2000, when nuclear-powered spaceships would be ferrying men to Mars (Seaborg 1972).

In reality, the 1980s saw an almost complete end of orders for new nuclear plants throughout the rich world. Fission's dismal commercial prospects were further damaged by the Chernobyl accident. Renewable energies in general, and solar conversions in particular, stepped into the mythical void created by fission's demise. But after two decades of intensive research and development neither photovoltaics nor other direct generation schemes have reached the stage of broad commercial acceptance.

What can be foreseen with great certainty is that much more energy will be needed during the coming generations to extend decent life to the still growing global population. This task may seem overwhelming or even impossible. Global high-energy civilization already suffers economically and socially from its precipitous expansion, and its further growth threatens the biospheric integrity on which its very survival depends.

Yet another great uncertainty is the long-term viability of urban living (McNeill 1993). The social cohesion and family nurturing so characteristic of rural life clearly do not prevail in modern cities. The strains of urban living are manifest in both rich and poor nations. Already many of the world's largest cities—perhaps most notably Los Angeles, New York, Ciudad Mexico, Rio de Janeiro, Cairo, New Delhi, and Bangkok—are epitomes of violence, drug addiction, homelessness, child abandonment, prostitution, and squalid living. And yet, perhaps more than ever, the imperatives of modern economies demand social stability, continuity, and effective cooperation. Cities have always been renewed by migration from villages—but what will happen to the already mostly urban civilization once the villages virtually disappear while the social structure of cities continues to disintegrate?

But there are hopeful signs. Precisely because gross energy use does not determine the course of history, humankind's commitment and inventiveness can go a
long way toward first weakening and then even reversing the evolutionary link
between civilization's advances and energy. We now realize that growing energy
use cannot be equated with effective adaptations and that we should be able to
stop and even reverse that trend, to break the dictum of Alfred Lotka's (1925) law
of maximum energy. This task should be easier once society realizes that it is
counterproductive to maximize power outputs.

Indeed, higher energy use by itself does not guarantee anything except greater
environmental burdens (Smil 1991). The historical evidence is clear. Higher energy use will not assure a reliable food supply (as the contrast between czarist
Russia and the Soviet Union demonstrates); it will not confer strategic security
(the United States was surely more secure in 1885 than in 1985); it will not safely
underpin political stability (be it in Brazil, Italy, or Egypt); it will not necessarily
lead to a more enlightened governance (it surely has not in North Korea or Iran);
and it will not automatically bring widely shared increases in standard of living
(not in Guatemala or Nigeria).

Opportunities for a grand transition to a less energy-intensive society can be
found primarily among the world's preeminent abusers of energy and materials
in Western Europe, North America, and Japan. Many of these savings could be
surprisingly easy to realize. I agree with George Basalla (1980) that the energy-civilization equation "should be exposed and discarded because it supplies a suppos-
edly scientific argument against efforts to adopt a style of living based upon lower
levels of energy consumption" (p. 51). Doing so would have profound consequences for assessing the prospects of high-energy civilization—but the chances
of success remain unknown.

Life's two cardinal characteristics have been expansion and increasing complexity. Can we reverse these trends by adopting the technically feasible and envi-
ronmentally desirable shift to moderated energy uses? Can we continue human
evolution by concentrating only on those aspects that do not require maximization of energy flows? Collectively, this shift would require the acceptance of low-
growth, and eventually even no-growth, economies. For individuals this would
mean a no less revolutionary delinking of social status from material consumption. Setting up such societies would be especially burdensome for the first generations making the transition. In the longer run, these new arrangements would
also eliminate one of the mainsprings of Western progress, the quest for social
and economic mobility.

The chances of succeeding in this unprecedented quest remain uncertain.
Given our degree of understanding, the challenge may not be any more forbid-
ding than overcoming a number of barriers we surmounted in the past. But understanding, no matter how impressive, will not be enough. What is needed is a
commitment to change, so we could say, paraphrasing Etienne Senancour (1804),
that if it is a failure that awaits us, let us not act so that it shall be a just fate.

Basic Measures

Length, mass, time, and temperature are the basic units of scientific accounts. The meter (m) is the basic unit of length. For average-sized people it is roughly the distance between their waist and the ground. Most people are between 1.5 and 1.8 m tall, ceilings of American houses are about 2.5 m high, Olympic track runs 400 m, a jet runway 3000 m. Standard (Greek) prefixes are used to express large values of scientific units (the full list appears at the end of this section). A kilo is 1000 and hence 3000 m are 3 kilometers (km). A coast-to-coast flight is 4000 km, the equatorial circumference about 40,000 km. Light travels 300,000 km every second, and 150 million km separate the Earth from the sun. Standard prefixes (Latin) are also used for fractional units (see list). A centimeter (cm) is a hundredth part of 1 meter. A fist resting on a table with the thumb alongside the bent fingers will be about 10 centimeters (one-tenth of a meter) long. New pencils measure about 20 cm (0.2 m), newborn babies 50 cm (0.5 m).

Area units are simply the squares of lengths. A coaster covers about 10 cm^2, a bed about 2 m^2, the foundations of a small bungalow about 100 m^2. This area (10 × 10 m) is called an are, and one hundred (hecto) such squares add up to a hectare (ha), the basic metric unit used for measuring agricultural land. Chinese or Bangladeshis must feed themselves from less than one-tenth of a hectare per person, but Americans cultivate nearly 1 hectare per capita. Outside agriculture, larger areas are usually expressed in square kilometers (km^2). North American cities of about 1 million people usually cover less than 500 km^2, small European countries have well below 100,000 km^2, and the United States takes up nearly 10 million km^2.

Basic mass units can be easily derived by filling cubes with water. A tiny cube enclosing 1 cubic centimeter (cm^3)—its side only as long as the width of a small fingernail—will weigh (or, more precisely, will have a mass) of 1 gram (g) when filled with water. A fist-sized cube will enclose 1000 cm^3 (10 × 10 × 10 cm), or 1 liter (l) of volume. When filled with water it will have a mass of 1 kilogram (kg). The kilogram is the basic unit of mass. Soft drinks weigh about one-third of 1 kg (350 g), newborn babies between 3 and 4 kg, and most adults between 50 and 100 kg. Compact cars have a mass around 1000 kg, or 1 tonne (t). A big horse will weigh as much, railway cars range from 30–100 t, ships from a few thousand to 500,000 t.

The second (s), a time span slightly shorter than an average heartbeat, is the basic unit of time. When at rest, we take a breath every 4 seconds, and it takes about 10 seconds to drink a glass of water. Larger time units are exceptions within the metric system of scientific units. They do not go up with multiples of 10. A red light at a busy intersection lasts 60 seconds, or 1 minute. Hardboiling an egg takes 8 minutes, an average classical symphony plays for 40 minutes. A normal pregnancy lasts 280 days, the average Western life span is 72 years, agriculture started to spread about 10,000 years ago, dinosaurs were plentiful 80 million years ago, and the Earth is 4.5 billion years old.

The scientific scale for temperature, degrees of Kelvin, starts at absolute 0. The Celsius scale divides the span between the freezing and boiling point of water into 100 degrees (°C). On that scale absolute 0 is –273.15 °C, water freezes at 0 °C, a fine spring day is around 20 °C, and the normal human temperature is 37 °C. Water boils at 100 °C, paper ignites at 230 °C, iron melts at 1535 °C, and the sun's thermonuclear reactions proceed at 15 million°C.

Nearly all other scientific units can be derived from length, mass, time, and temperature. For energy and power the derivations are as follows. A force acting on a mass of 1 kg having an acceleration of 1 m/s^2 is equal to one newton (N). The force of 1 N applied over a distance of 1 m equals 1 joule (J), the basic unit of energy. As power is the rate of energy use, one J/s is equal to 1 watt (W).

Many units, such as those for speed—meters per second (m/s) or kilometers per hour (km/h)—and productivity—kilograms or tonnes per hour (kg/h; t/h) or tonnes per year (t/y)—have no special name. Working horses move at about 1 m/s, most highway speed limits are around 100 km/h. A slave milling grain with a stone quern produced flour at a rate no higher than 4 kg/h, an excellent crop of late medieval wheat was 1 t/ha.

Prefixes of Scientific Units

Prefixes used to express decimal submultiples (one-tenth, one-hundredth, etc.) have specific names for the first three orders of magnitude and then only for every third one.

Common name	Prefix	Symbol	Scientific Notation
one-tenth	deci	d	10^{-1}
one-hundredth	centi	c	10^{-2}
one-thousandth	milli	m	10^{-3}
one-millionth	micro	μ	10^{-6}

Prefixes for multiples also have special names for the first three orders of magnitude and then only for every thousandfold increase.

Common name	Prefix	Symbol	Scientific Notation
ten	deca	da	10^{1}
hundred	hecto	h	10^{2}
thousand	kilo	k	10^{3}
million	mega	M	10^{6}
billion	giga	G	10^{9}
trillion	tera	T	10^{12}
quadrillion	peta	P	10^{15}
quintillion	exa	E	10^{18}

Chronology of Energy-Related Developments

Space considerations restrict this list almost completely to practical technical advances. Therefore, the list excludes the underlying intellectual, scientific, and political developments, and it contains only a few economic and environmental milestones. All early dates are approximations, and other sources may list different ones. Discrepancies exist even with modern advances: Dates may refer to the the year when an item was originally conceived or invented, to patenting, to the first practical application, or to successful marketing. For problems in dating inventions see Petroski (1993).

B.C.

1500000+	Oldowan stone tools (< 0.5 m of edge/kg of stone)
600000+	Abbevillian stone tools (1 m of edge/kg stone)
250000+	Acheulean stone tools (3 m of edge/kg of stone)
130000+	Mousterian stone flake tools (4 m of edge/kg of stone)
50000+	Bone objects
30000+	Aurignacian stone tools (10 m of edge/kg of stone)
	Bow and stone arrows
15000+	Magdalenian stone tools (12 m of edge/kg of stone)
9000+	Sheep domesticated in the Middle East
7400+	Corn in Oaxaca Valley
7000+	Wheat in Mesopotamia
	Pigs domesticated in the Middle East
6500+	Cattle domesticated in the Middle East
6000+	Copper artifacts more common in the Middle East
5000+	Barley in Egypt
	Corn in the Basin of Mexico
4400+	Potatoes in highland Peru and Bolivia
4000+	Light wooden plows in Mesopotamia
	Wooden ships (Mediterranean, Indian Ocean)
3500+	Pack asses in the Middle East
	Pottery and bricks fired in kilns in Mesopotamia
	Irrigation in Mesopotamia
3200+	Wheeled vehicles in Uruk

3000+	Square sails in Egypt
	Draft oxen in Mesopotamia
	Camels domesticated
	Potter's wheels in Mesopotamia
2800+	Pyramid construction in Egypt
2500+	Bronze in Mesopotamia
	Small glass objects in Egypt
2000+	Spoked wheels in Mesopotamia
	Horse-drawn vehicles in Egypt
	Shaduf in Mesopotamia
1700+	Horse riding on Euro-Asian steppes
1500+	Copper in China
	Paddy rice in China
	Axle lubricants in the Middle East
1400+	Iron in the Middle East
1300+	Seed drills in Mesopotamia
	Horse-drawn chariots in China
1200+	Iron more common in India, the Middle East, Europe
800+	Mounted archers on Asian steppes
	Candles in the Middle East
600+	Tin in Greece
	Penteconter ships common in Greece
	Archimedean screws in Egyptian irrigation
500+	Camel saddles in North Arabia
	Triremes in Greece
400+	Crossbows in China
432	Parthenon completed
300+	Stirrups in China
	Gears in Egypt and Greece
312	Roman *Via Appia* and *Aqua Appia* completed
200+	Breastband harnesses in China
	Sailings to windward advances in China
	Batten-strengthened sails in China
	Percussion drilling in Sichuan
	Crank handles in China
150+	Iron moldboard plows in China
100+	Beginnings of collar harness in China
	House heating by coal in China
	Waterwheels in Greece and Rome
	Wheelbarrows in China
	Norias in the Middle East
80	*Hypocaust* heating in Rome

A.D.

600+	Windmills in Iran
850+	Triangular sails in the Mediterranean
900+	Collar harnesses and horseshoes common in Europe
	Bamboo fire lances in China
980s	Canal pound locks in China
1000+	Waterwheels widespread in Western Europe
1040	Clear directions for gunpowder preparation in China
1100+	Long bows in England
1150+	Windmills spread in Western Europe
	Cathedrals in Europe
1200+	Incas road construction
1280+	Cannons in China
1300+	Gunpowder and cannons in Europe
1327	Beijing-Hangzhou Grand Canal (1800 km) completed
1350+	Handheld guns in Europe
1400+	Heavy draft horses in Europe
	Drainage windmills in the Netherlands
	Blast furnaces in the Rhine region
1420+	Portuguese caravels make longer sailings
1492	Columbus sails across the Atlantic
1497	Vasco da Gama sails to India
1519	Magellan's *Victoria* circumnavigates the Earth
1550+	Large full-rigged sailships with guns in Western Europe
1600+	Ball bearings in Western Europe
1640+	English coal mining expands
1690	Experiments with atmospheric steam engine (Denis Papin)
1698	Simple, small steam engines (Thomas Savery)
1709	Coke from bituminous coal (Abraham Darby)
1712	Atmospheric steam engines (Thomas Newcomen)
1745	Fantails for automatic turning of windmills
1750+	Intensive canal construction in Western Europe
	Use of coke spreads in English ironmaking
	Newcomen's engine more common in English coal mines
1757	Precision-cutting lathes (Henry Maudslay)
1769	Separate condensers for steam engines (James Watt)
1770s	Factories powered by waterwheels
1775	Watt's patent on steam engines extended to 1800
1782	Hot air balloons (Joseph and Etienne Montgolfier)

1794	Lamps with wick holders and glass chimneys (Aime Argand)
1800s	Steamboats (*Charlotte Dundas, Clermont*)
	High-pressure steam engines (R. Trevithick, O.Evans)
1800	Electric batteries (Alessandro Volta)
1805	Steam-powered cranes (J. Rennie)
	Coal (town) gas in England
1808	Arc lamps (Humphrey Davy)
1809	Chilean nitrates discovered
1816	Mine safety lamps (Humphrey Davy)
1820s	Designs of mechanical calculators (Charles Babbage)
	Iron ship hulls
1820	Electromagnetism (Hans C. Oersted)
1823	Silicon isolated (J. J. Berzelius)
1824	Portland cement (Joseph Aspdin)
	Aluminum isolated (Hans C. Oersted)
1825	Stockton-Darlington railway
1828	Hot blast in ironmaking (James Neilson)
1829	*Rocket* locomotive (Robert Stephenson)
1830s	Railway construction takes off in England
	Steamships cross the Atlantic
	Mechanical grain reapers (Cyrus McCormick, Obed Hussey)
1830	Thermostats (Andrew Ure)
	Liverpool-Manchester railway
1832	Water turbines (Benoit Fourneyron)
1833	Steel plows (John Lane)
	Steamship *Royal William* crosses from Quebec to London
1834	Free-standing kitchen ranges (Philo P. Stewart)
1837	Electric telegraph (W. Cooke, C. Wheatstone)
1838	Screw propulsion for steamships (John Ericsson)
	Telegraph code (Samuel Morse)
1840s	Peak decade of U.S. whaling
1841	Steam-driven threshing machines
	Thomas Cook offers holiday trips
1847	Inward-flow water turbines (James B. Francis)
1850s	Paraffin from oil for lighting
	Fast clippers on long voyages
1852	Hydrogen-filled airship (Henri Giffard)
1854	*Great Eastern* steamship (Isambard K. Brunel)
1856	Steel convertors (Henry Bessemer)
1858	Grain harvesters (C. W. and W. W. Marsh)

1859	Crude oil discovered in Pennsylvania (E. L. Drake)
1860s	Steam plowing of large U.S. fields
1860	Horizontal internal combustion engines (J.J.E. Lenoir)
	Milking machines (L. O. Colvin)
1864	Open-hearth steel-making process (W. and F. Siemens)
1865	Nitrocellulose (J.F.E. Schultze)
1866	Carbon-zinc batteries (Georges Leclanche)
	Transatlantic cable in permanent operation
	Torpedoes (Robert Whitehead)
1867	Refrigerated railway wagons in service
1869	Suez Canal completed
	U.S. transcontinental railroad completed
1870s	Refrigerated transport of meat by ocean ships
	Phosphate fertilizer industry begins
1871	Ring-wound armature dynamo (Z. T. Gramme)
1875	Dynamite (Alfred Nobel)
1874	Photographic film (G. Eastman)
1876	Four-stroke internal combustion engines (N. A. Otto)
	Telephone patented (Alexander Graham Bell, Elisha Gray)
1877	Phonographs (Thomas A. Edison)
1878	Two-stroke internal combustion engines (Dugald Clerk)
	Filament light bulbs (Joseph Swan)
	Twine knotters for grain harvesters (John Appleby)
1879	Carbon filament light bulbs (Thomas A. Edison)
1880s	Horse-drawn grain combines (California)
	Modern bicycles (J. K. Starley, William Sutton)
	Crude oil tankers
	Military high explosives formulated
1882	Edison's first electricity-generating plants
1883	Impulse steam turbines (Carl Gustaf de Laval)
	Four-stroke liquid-fueled engines (Gottlieb Daimler)
	Machine guns (H. S. Maxim)
1884	Steam turbines (Charles Parsons)
1885	Transformers (William Stanley)
	Karl Benz builds the first practical car
1886	Prestressed concrete (C. E. Dochring)
	Aluminum production (C. M. Hall, P.L.T. Heroult)
1887	Crude oil discovered in Texas
	Generation of electromagnetic waves (Heinrich Hertz)
1888	Induction electric motors (Nikola Tesla)
	Gramophones (Emile Berliner)
	Air-filled rubber tires (John B. Dunlop)

1889	Cylindrical phonographs (Thomas A. Edison)
	Jet-driven water turbines (Lester A. Pelton)
1890s	Horses reach peak numbers in Western cities
	Electric household appliances introduced
1892	Diesel engines (Rudolf Diesel)
1894	Offshore oil drilling from jetties (California)
1895	Moving pictures (Louis and August Lumiere)
	X-rays (W. K. Roentgen)
1897	Cathode-ray tubes (Ferdinand Braun)
1898	Tape recorders (Valdemar Poulsen)
1899	Radio signals across the Channel (Guglielmo Marconi)
1900s	Electricity consumption takes off in the United States and the United Kingdom
	Mass production of cars begins
1900	Dirigible powered airship (Ferdinand von Zeppelin)
1901	Industrial air conditioning (Willis H. Carrier)
	Rotary drilling (Spindletop, Texas)
	Radio signals across the Atlantic (Guglielmo Marconi)
1903	Sustained controlled powered flight (Orville and Wilbur Wright)
1904	Geothermal electricity generation (Lardarello, Italy)
	Vacuum diodes (John A. Fleming)
1905	Photoelectric cells (Arthur Korn)
	Commercial tractor production in the United States
1906	British *Dreadnought* battleship launched
	Vacuum triodes (Lee De Forest)
1908	Tungsten light bulbs
	Ford's Model T (until 1927)
1909	Rolling cutter rock drilling bits (Howard Hughes)
	Louis Bleriot flies across the Channel
	Bakelite, the first major plastic (Leo Baekeland)
1910	Neon light (Georges Claude)
	Synthetic gas from coal (Fischer-Tropsch, Germany)
1913	Moving production line (Ford Company)
	Panama Canal completed
	Ammonia synthesis (Fritz Haber, Carl Bosch)
	High-pressure crude oil cracking (W. M. Burton)
1916	Tanks in battle
1919	Nonstop transatlantic flight (J. Alcock, A. W. Brown)
	Scheduled airline service (Paris-London)
1920s	Boilers burning pulverized coal
	Streamlined metal plane bodies
	Electric record players

Radio broadcasting spreads in North America and Europe

Liquefaction of coal (Friedrich Bergius)

1920	Axial-flow water turbines (Viktor Kaplan)
1922	Aircraft carrier *Hosho* launched in Japan
1923	Electronic camera tubes (Vladimir Zworykin)
	Electric refrigerators by Electrolux
1927	Synthetic rubber (Buna)
	Nonstop solo transatlantic flight (C. A. Lindbergh)
1928	Plexiglass (W. Bauer)
1929	Experimental TV broadcasts (United Kingdom)
1930s	Catalytical crude oil cracking (Eugene Houdry)
	Large hydrostations (United States and Soviet Union)
	Long-distance bombers
	Chlorofluorocarbons in refrigeration
1933	Polyethylene (Imperial Chemical Industries)
1935	Fluorescent lights (General Electric)
	Plastic magnetic tape (AEG Telefunken, I. G. Farben)
	Nylon (Wallace Carothers)
1936	Regular TV broadcasting (BBC)
	Gas turbines (Brown-Boveri)
1937	Fully pressurized aircraft (Lockheed XC-35)
1938	Prototype jet fighter (Hans Pabst von Ohain)
1940s	Military jet aircraft
	Radar
	Electronic computers
1940	Helicopters (Igor Sikorsky)
1941	Pearl Harbor attacked by Japanese carrier planes
1942	V-1 rockets (Wernher von Braun)
	Industrial production of silicone
	Controlled chain reaction (Enrico Fermi, Chicago)
1944	V-2 rockets
	DDT marketed
1945	Nuclear bombs (Trinity test, Hiroshima and Nagasaki)
	Electronic computer (ENIAC, United States)
	The first herbicide (2,4-D) marketed
1947	Transistors (J. Bardeen, W. H. Brattan, W. B. Shockley)
	Offshore drilling out of sight of land (Louisiana)
	Piloted flight faster than the speed of sound (Bell X-1)
1948	Basic-oxygen steelmaking furnace (Linz-Donawitz)
	The world's largest oilfield, Saudi al-Ghawar, drilled
1949	Passenger jet aircraft (De Havilland Comet)

1950s	Rapidly growing worldwide crude oil consumption
	Continuous pig iron casting
	Electrostatic precipitators spread
	Commercial computers
	Stereo recordings
	Videotape recorders
1951	Automatic engine assembly (Ford Company)
	Transmission of color TV images
	Hydrogen (fusion) bomb
1952	British Comet jetliner in commercial service
1953	Microwave oven (Raytheon Manufacturing Company)
1954	U.S. Navy nuclear submarine *Nautilus* launched
1955	Fully transistorized Sony radio
1956	First commercial nuclear power plant (Calder Hall, United Kingdom)
	Transatlantic telephone cable
	Interstate highway construction begins in the United States
1957	*Sputnik,* the first Soviet satellite
	First U.S. nuclear power plant (Shippingport, Pennsylvania)
1958	Integrated circuits (Texas Instruments)
	U.S. Boeing 707 jetliner in service
1960s	Semisubmersible platforms for offshore drilling
	Meteorological and communications satellites
	Very large crude oil tankers
	Large-scale deployment of intercontinental ballistic missiles (ICBMs)
	Largest Soviet fusion bombs tested in the atmosphere
	Spreading use of synthetic fertilizers and pesticides
	High-yielding crop varieties
	Concerns about environmental pollution in the West
1960	U.S. *Minuteman* ICBM tested
1961	U.S. nuclear-powered aircraft carrier *Enterprise* launched
	Manned space flight (Yuri Gagarin)
1962	Transatlantic television relay (Telstar)
1964	*Shinkansen,* Japan National Railway's superexpress
1965	Liquefied natural gas exports from Algeria
1966	U.S. jumbo jet aircraft Boeing 747 takes off
1969	British-French supersonic aircraft Concorde takes off
	Boeing 747 in commercial service

	Microprocessors (Edward Hoff, Intel)
	U.S. *Apollo* spacecraft lands on the moon
1970s	Radio and television satellite broadcasting
	Acid rain over Europe and North America
	Energy prices rise after decades of decline
	Interest in renewable energy conversions
1971	Silicon chips (Intel and Texas Instruments)
1973	OPEC's first round of crude oil price increases
1976	Concorde in commercial service
	Unmanned U.S. *Voyager* spacecraft lands on Mars
1977	Man-powered flight of *Gossamer Condor*
1979	OPEC's second round of crude oil price increases
	Three Mile Island nuclear power plant accident
1980s	Personal computers take off
	Concerns about global environmental change mount
1982	CD players (Philips, Sony)
1983	TGV, the French superexpress train
1985	World oil prices decline rapidly
1986	Chernobyl nuclear reactor disaster
1990	Global population surpasses 5 billion

Compiled from a wide variety of sources cited in the References. More extensive chronologies of technical advances can be found in Mumford (1934), Gille (1978), Taylor (1982), Temple (1986), and Williams (1987).

Power in History

Actions, Prime Movers, Convertors	Power (W)
Small wax candle burning (800 B.C.)	5
Egyptian boy turning an Archimedean screw (500 B.C.)	25
Small U.S. windmill rotating (1880)	30
Chinese woman cranking a winnowing machine (100 B.C.)	50
French glass polishers working steadily (1700)	75
Strong man rapidly treading a wooden wheel (1400)	200
IBM personal computer used to write this book (1993)	287
Donkey turning a Roman hourglass mill (100 B.C.)	300
Weak pair of Chinese oxen plowing (A.D. 200)	600
Good English horse turning a whim (1770)	750
Eight men powering a Dutch treadwheel (1500)	800
Very strong American horse pulling a wagon (1890)	1000
Long-distance runner competing at the Olympic games (600 B.C.)	1400
Roman vertical waterwheel turning a millstone (A.D. 50) 1800	
Newcomen's atmospheric engine pumping water (1712)	3750
Engine of Ransom Olds's *Curved Dash* automobile (1904)	5200
Fifty oarsmen powering a Greek *penteconter* at full speed (600 B.C.)	6000
Large German post windmill crushing oilseeds (1500)	6500
Roman messenger's horse galloping (A.D. 200)	7200
Large Dutch windmill draining a polder (1750)	12,000
Engine of a Ford Model T running at full speed (1908)	14,900
One hundred and seventy oarsmen powering a Greek *trireme* at full speed (500 B.C.)	20,000
Watt's steam engine winding coal (1795)	20,000
Team of forty horses pulling a California combine (1885)	28,000
Cascade of sixteen Roman watermills at Barbegal (A.D. 350)	30,000
Benoit Fourneyron's first water turbine (1832)	38,000
Water pumps for Versailles at Marly (1685)	60,000
Engine of Honda Civic GL (1985)	63,000
Charles Parsons's steam turbine (1888)	75,000
Steam engine at Edison's Pearl Street Station (1882)	93,200
Watt's largest steam engine (1800)	100,000

Electricity used by a U.S. supermarket (1980)	200,000
Diesel engine of a German submarine (1916)	400,000
Lady Isabella, the world's largest waterwheel (1854)	427,000
Large steam locomotive at full speed (1890)	850,000
Parsons's steam turbine at Elberfeld station (1900)	1,000,000
Shaw's water works at Greenock, Scotland (1840)	1,500,000
Rocket engine launching a V-2 missile (1944)	6,200,000
Gas turbine powering a pipeline compressor (1970)	10,000,000
Japanese merchant ship's diesel engine (1960)	30,000,000
Four gas turbines of a Boeing 747 (1969)	60,000,000
Calder Hall nuclear reactor (1956)	202,000,000
Turbogenerator at Chooz nuclear station (1990)	1,457,000,000
Rocket engines launching the Saturn C5 (1969)	2,600,000,000
Fukushima nuclear power station (1990)	9,096,000,000
U.S. coal and biomass energy consumption (1850)	79,000,000,000
U.S. commercial energy consumption (1990)	2,300,000,000,000
Global commercial energy consumption (1990)	9,500,000,000,000

Suggested Readings

Advances in energy use are described systematically in the outstanding multivolume histories of technical progress by Singer et al. (1954–1958), Forbes (1964–1972), and Needham et al. (1954–). Energy matters are covered with varying degrees of detail in many writings tracing the history of inventions and engineering practices. Perhaps the most notable among these publications are books by Byrn (1900), Neuburger (1930), Abbott (1932), Mumford (1934), Usher (1954), Klemm (1964), de Camp (1960), Derry and Williams (1960), Burstall (1963), Daumas (1969), Lindsay (1974), Gille (1978), L. White (1978), Landels (1980), Taylor (1982), Hill (1984), K. D. White (1984), Williams (1987), Basalla (1988), Hodges (1990), Pacey (1990), and Finniston et al. (1992).

The prime movers in evolutionary sequence, human energetics, food consumption, and growth of global population are covered in Snyder (1930), Durnin and Passmore (1967), McKeown (1976), Chandler (1987), McEvedy and Jones (1978), FAO/WHO/UNU (1985), Anderson (1988), Baines (1991), Livi-Bacci (1991), and FAO (1992). For information about the contributions of horses to civilization, consult des Noettes (1931), Smythe (1967), Dent (1974), Silver (1976), Villiers (1976), Telleen (1977), Baskett (1980), Langdon (1986), Berkebile (1989), and Hyland (1990). Other draft and pack animals are dealt with in Hopfen (1969), Rouse (1970), Cockrill (1974), Bulliet (1975), and Choudhury (1976).

The long history of waterwheels and their importance during the time of early industrialization can be traced in volumes by Wilson (1956), Forbes (1958; 1965), Hindle (1975), Fox (1976), Priamo (1976), and White (1978). The history of windmills and their economic importance are well reviewed in Wolff (1900), Skilton (1947), Freese (1957), Stockhuyzen (1963), Needham et al. (1965), Reynolds (1970), and Torrey (1976). The development of sailing ships is traced comprehensively in Torr (1964), Armstrong (1967; 1969), Barjot and Savant (1965), and Chatterton (1926; 1977). Interesting volumes on oared ships include Lane (1934), Morrison and Williams (1968), and Morrison and Coates (1986).

Indispensable sources for the history of steam engines are Farey (1827), Dalby (1920), Dickinson (1939; 1967), Watkins (1967), Robinson and Musson (1969), Jones (1973), and von Tunzelmann (1978). Reviews by Saxena (1962), Haney (1968), Ellis (1977), and O'Brien (1983) are just a few notable items among the rich literature on steam locomotives and railways. Advances in steamships are recounted by Fry (1896), Croil (1898), and Wood (1922).

The development of internal combustion engines and gas turbines is reviewed in Williams (1972), Constant (1981), Taylor (1984), and Gunston (1986). The automobile age is chronicled by Flower and Jones (1981), Flink (1988), Ling (1990), and Womack et al. (1991). The history of flying is recounted by Wright (1953), Taylor (1989), and Jakab (1990). The pioneering decades of the electrical industry and its subsequent expansion are traced by Jehl (1937), Lilienthal (1944), Josephson (1959), Electricity Council (1973), Cheney (1981), Hughes (1983), and Nye (1992).

The properties and uses of biomass energies are covered in Earl (1973), Tillman (1978), Hall et al. (1982), and Smil (1983). The discoveries, development, and uses of fossil fuels are detailed in Ashton and Sykes (1929), Nef (1932), Eavenson (1942), Brantly (1971), and Perrodon (1985).

The literature on the history of productive human activities is very rich. Perspectives on agricultural development, from farming's origins to the twentieth century, can be found in Bailey (1908), King (1927), Seebohm (1927), Buck (1930; 1937), Leser (1931), Lizerand (1942), Haudricourt and Delamarre (1955), Geertz (1963), Slicher van Bath (1963), Allan (1965), Boserup (1965), Perkins (1969), Titow (1969), Clark and Haswell (1970), White (1970), Fussell (1972), Ho (1975), Schlebecker (1975), Cohen (1977), Reed (1977), Abel (1980), Jarman et al. (1982), Bray (1984), Rindos (1984), and Cowan and Watson (1992). Details on water lifting and irrigation are contained in Ewbank (1870), Needham et al. (1965), Butzer (1976), Paul (1983), Oleson (1984), Fraenkel (1986), and Molenaar (1956). The energy costs of modern agriculture are exhaustively reviewed in Pimentel (1980), Helsel (1987), and Fluck (1992a).

Interdisciplinary insights into the origins, process, and consequences of industrialization can be found in Kay (1832), Clapham (1926), Ashton (1948), Landes (1961), Falkus (1972), Mokyr (1976), Boustead and Hancock (1979), Clarkson (1985), Rosenberg and Birdzell (1986), and Blumer (1990). Many aspects of construction activities are chronicled and explained by Ashby (1935), Fitchen (1961), Niel (1961), Bandaranayke (1974), Baldwin (1977), Bagenal and Meades (1980), Hodges (1989), Lepre (1990), Waldron (1990), and Wilson (1990). Contributions to the history of transportation include books by Hadfield (1969), Sitwell (1981), Piggott (1983), Ratcliffe (1985), and Ville (1990).

Metallurgical progress can be traced in books by Biringuccio (1540), Agricola (1556), Bell (1884), Campbell (1907), Greenwood (1907), Boylston (1936), King (1948), Needham (1964), Straker (1969), Hogan (1971), Hyde (1977), Peacey (1979), Gold et al. (1984), Haaland and Shinnie (1985), and Harris (1988). Weapons from ancient to modern times and their effects on societies are reviewed in Mitchell (1931), Liddell Hart (1954; 1970), Kloss (1963), Cipolla (1966), Ziemke (1968), Egg et al. (1971), Singer and Small (1972), Craig (1973), von Braun and Ordway (1975), Kesaris (1977), Anderson (1988), McNeill (1989), and DataCenter (1991).

Writings on the broad social implications of energy use in civilization include books by Ellul (1964), Jones (1970), Odum (1971), Adams (1975; 1982), Cook (1976), and Smil (1991). Long-term economic trends are analyzed in Schumpeter (1939), Rostow (1965), Kuznets (1971), van Duijn (1983), Freeman (1983), and Vasko et al. (1990). Finally, anybody who wants to study history through the evolution of tools and machines must consult books that are appropriately illustrated. The two unsurpassed classical works are Ramelli (1588) and Diderot and D'Alembert (1769–1772). Ardrey (1894), Abbott (1932), Hommel (1937), Burstall (1968), Hopfen (1969), Taylor (1982), Williams (1987), Basalla (1988), and Finniston et al. (1992) are the notable modern contributions.

References

Abbott, C. G. 1932. *Great Inventions*. Smithsonian Institution, Washington, D.C.

Abel, W. 1980. *Agricultural Fluctuations in Europe*. St. Martin's Press, New York.

Adams, R. N. 1975. *Energy and Structure: A Theory of Social Power*. University of Texas Press, Austin.

————. 1982. *Paradoxical Harvest*. Cambridge University Press, Cambridge.

Adshead, S.A.M. 1992. *Salt and Civilization*. St. Martin's Press, New York.

Agricola, G. 1556. *De re metallica*. Translated by H. C. Hoover and L. H. Hoover. Dover Publications, New York (1950).

Allan, W. 1965. *The African Husbandman*. Oliver & Boyd, Edinburgh.

Anderson, E. N. 1988. *The Food of China*. Yale University Press, New Haven.

Anderson, M. S. 1988. *War and Society in Europe of the Old Regime, 1618–1789*. St. Martin's Press, New York.

Anthony, D., D. Y. Telegin, and D. Brown. 1991. "The Origin of Horseback Riding." *Scientific American* 265(6):94–100.

Ardrey, R. L. 1894. *American Agricultural Implements*. Published by the author, Chicago.

Armstrong, R. 1967. *The Early Mariners*. Ernest Benn, London.

————. 1969. *The Merchantmen*. Ernest Benn, London.

Ashby, T. 1935. *The Aqueducts of Ancient Rome*. Oxford University Press, Oxford.

Ashton, B., K. Hill, A. Piazza, and R. Zeitz. 1984. "Famine in China, 1958–61." *Population and Development Review* 10:613–645.

Ashton, T. S. 1948. *The Industrial Revolution, 1760–1830*. Oxford University Press, London.

Ashton, T. S., and J. Sykes. 1929. *The Coal Industry of the 18th Century*. Manchester University Press, Manchester.

Ayres, J. C., O. Mundt, and W. E. Sandine. 1980. *Microbiology of Foods*. W. H. Freeman, San Francisco.

Ayres, R. U. 1969. *Technological Forecasting and Long-Range Planning*. McGraw-Hill, New York.

Baars, C. 1973. *De Geschiedenis van de Landbouw in de Bayerlanden*. PUDOC, Wageningen.

Bagenal, P., and J. Meades. 1980. *Great Buildings*. Galahad Books, New York.

Bailey, L. H., ed. 1908. *Cyclopedia of American Agriculture*. Macmillan, New York.

Bailey, R. C., G. Head, M. Jenike, B. Owen, R. Rechtman, and E. Zechenter. 1989. "Hunting and Gathering in Tropical Rain Forest: Is It Possible?" *American Anthropologist* 91:59–82.

Baines, D. 1991. *Emigration from Europe 1815–1930*. Macmillan, London.

Bairoch, P. 1991. "The City and Technological Innovation." In P. Higonnet, ed., *Favorites of Fortune*, Harvard University Press, Cambridge, Mass., pp. 159–176.

Baldwin, G. C. 1977. *Pyramids of the New World*. G. P. Putnam's Sons, New York.

Bandaranayake, S. 1974. *Sinhalese Monastic Architecture*. E. J. Brill, Leiden.

Barber, F. M. 1900. *The Mechanical Triumphs of Ancient Egyptians.* Tribner, London.

Barjot, A., and J. Savant. 1965. *History of the World's Shipping.* Ian Allan, London.

Basalla, G. 1980. "Energy and Civilization." In C. Starr and P. C. Ritterbush, eds., *Science, Technology and the Human Prospects,* Pergamon Press, New York, pp. 39–52.

———. 1982. "Some Persistent Energy Myths." In G. H. Daniels and M. H. Rose, eds., *Energy and Transport,* Sage Publications, Beverly Hills, Calif., pp. 27–38.

———. 1988. *The Evolution of Technology.* Cambridge University Press, Cambridge.

Baskett, J. 1980. *The Horse in Art.* New York Graphic Society, Boston.

Bell, L. 1884. *Principles of the Manufacture of Iron and Steel.* George Routledge & Sons, London.

Bennett, M. K. 1935. "British Wheat Yield per Acre for Seven Centuries." *Economic History* 3(10):12–29.

Bennett, R., and J. Elton. 1898. *History of Corn Milling.* Simpkin Marshall, London.

Beresford, M. W., and J. G. Hurst. 1971. *Deserted Medieval Villages.* Littleworth, London.

Berkebile, D. H. 1989. *Horse-drawn Commercial Vehicles.* Dover, New York.

Bessemer, H. 1905. *Sir Henry Bessemer, F.R.S.* Offices of *Engineering,* London.

Bettinger, R. L. 1991. *Hunter-Gatherers: Archaeological and Evolutionary Theory.* Plenum Press, New York.

Biringuccio, V. 1540. *Pirotechnia.* Translated by C. S. Smith and M. T. Gnudi. Dover Publications, New York (1990).

Birt-David, N. 1992. "Beyond 'The Original Affluent Society.'" *Current Anthropology* 33:25–47.

Blumenschine, R. J., and J. A. Cavallo. 1992. "Scavenging and Human Evolution." *Scientific American* 267(4):90–95.

Blumer, H. 1990. *Industrialization as an Agent of Social Change.* Aldine de Gruyter, New York.

Boden, T. A., P. Kanciruk, and M. P. Farrell. 1990. *Trends '90.* Carbon Dioxide Information Analysis Center, Oak Ridge, Tenn.

Boserup, E. 1965. *The Conditions of Agricultural Growth: The Economics of Agrarian Change Under Population Pressure.* Aldine, Chicago.

———. 1976. "Environment, Population, and Technology in Primitive Societies." *Population and Development Review* 2(1):21–36.

Boulding, K. E. 1974. "The Social System and the Energy Crisis." *Science* 184:255–257.

Boustead, I., and G. F. Hancock. 1979. *Handbook of Industrial Energy Analysis.* Ellis Horwood, Chichester.

Boylston, H. M. 1936. *An Introduction to the Metallurgy of Iron and Steel.* John Wiley, New York.

Brantly, J. E. 1971. *History of Oil Well Drilling.* Gulf Publishing, Houston.

Braudel, F. 1980. "The History of Civilizations." In F. Braudel, *On History,* University of Chicago Press, Chicago, pp. 177–218.

Bray, F. 1984. *Science and Civilisation in China. Volume 6, Part II: Agriculture.* Cambridge University Press, Cambridge.

Bray, W. 1968. *Everyday Life of the Aztecs.* B. T. Batsford, London.

———. 1977. "From Foraging to Farming in Early Mexico." In J.V.S. Megaw, ed., *Hunters, Gatherers and First Farmers Beyond Europe,* Leicester University Press, Leicester, pp. 225–250.

Briggle. L. W. 1980. "Origin and Botany of Wheat." In *Wheat,* CIBA-Geigy, Basle, 6–13.

Brody, S. 1945. *Bioenergetics and Growth.* Reinhold, New York.

Brooks, D. R., and E. O. Wiley. 1986. *Evolution as Entropy.* University of Chicago Press, Chicago.

Buck, J. L. 1930. *Chinese Farm Economy.* University of Nanking, Nanking.

———. 1937. *Land Utilization in China.* University of Nanking, Nanking.

Bulliet, R. W. 1975. *The Camel and the Wheel.* Harvard University Press, Cambridge, Mass.

Burstall, A. F. 1963. *A History of Mechanical Engineering.* Faber & Faber, London.

———. 1968. *Simple Working Models of Historic Machines.* MIT Press, Cambridge, Mass.

Butzer, K. W. 1976. *Early Hydraulic Civilization in Egypt.* University of Chicago Press, Chicago.

———. 1984. "Long-term Nile Flood Variation and Political Discontinuities in Pharaonic Egypt." In J. D. Clark and S. A. Brandt, eds., *From Hunters to Farmers: The Causes and Consequences of Food Production in Africa,* University of California Press, Berkeley, Calif., pp. 102–112.

Byrn, E. W. 1900. *The Process of Invention in the Nineteenth Century.* Russell & Russell, New York.

Calder, W. A. 1983. "Ecological Scaling: Mammals and Birds." *Annual Review of Ecology and Systematics* 14:213–230.

Caldwell, J. C. 1976. "Toward a Restatement of Demographic Transition Theory." *Population and Development Review* 2:321–366.

Cameron, R. 1985. "A New View of European Industrialization." *The Economic History Review* 38:1–23.

Campbell, H. R. 1907. *The Manufacture and Properties of Iron and Steel.* Hill Publishing, New York.

Cardwell, D.S.L. 1971. *From Watt to Clausius.* Cornell University Press, Ithaca, N.Y.

Caro, R. A. 1982. *The Years of Lyndon Johnson: The Path to Power.* Knopf, New York.

Carrier, D. R. 1984. "The Energetic Paradox of Human Running and Hominid Evolution." *Current Anthropology* 25:483–495.

Carter, W. E. 1969. *New Lands and Old Traditions: Kekchi Cultivators in the Guatemala Lowlands.* University of Florida Press, Gainesville.

Centre des Recherches Historiques. 1965. *Villages Desertes et Histoire Economique.* SEVPEN, Paris.

Chandler, T. 1987. *Four Thousand Years of Urban Growth.* E. Mellen, Lewiston, N.Y.

Chatterton, E. K. 1926. *The Ship Under Sail.* Fisher Unwin, London.

———. 1977. *Sailing Ships: The Story of Their Development from the Earliest Times to the Present.* Gordon Press, New York.

Chen, A. 1990. "The Incredible Shrinking Transistor." *Microcomputer Solutions,* September/October:3–5.

Cheney, M. 1981. *Tesla: Man out of Time.* Prentice-Hall, Engelwood Cliffs, N.J.

Chi Ch'ao-ting. 1936. *Key Economic Areas in Chinese History, as Revealed in the Development of Public Works for Water Control.* Allen & Unwin, London.

Childe, V. G. 1951. "The Neolithic Revolution." In V. G. Childe, ed., *Man Makes Himself,* C. A. Watts, London, pp. 67–72.

Choudhury, P. C. 1976. *Hastividyarnava.* Publication Board of Assam, Gauhati.

Christ, K. 1984. *The Romans.* University of California Press, Berkeley.

CIA. 1980–. *Handbook of Economic Statistics.* CIA, Washington, D.C.

Cipolla, C. M. 1966. *Guns, Sails and Empires.* Pantheon Books, New York.

Clapham, J. H. 1926. *An Economic History of Modern Britain.* Cambridge University Press, Cambridge.

Clark, C., and M. Haswell. 1970. *The Economics of Subsistence Agriculture.* Macmillan, London.

Clark, G. 1987. "Productivity Growth Without Technical Change in European Agriculture Before 1850." *The Journal of Economic History* 47:419–432.

———. 1991. "Yields per Acre in English Agriculture, 1250–1850: Evidence from Labour Inputs." *Economic History Review* 44:445–460.

Clarke, R. A. 1972. *Soviet Economic Facts.* Macmillan, London.

Clarkson, L. A. 1985. *Proto-Industrialization: The First Phase of Industrialization?* Macmillan, London.

Coates, J. F. 1989. "The Trireme Sails Again." *Scientific American* 260(4):96–103.

Cockrill, W. R., ed. 1974. *The Husbandry and Health of the Domestic Buffalo.* FAO, Rome.

Cohen, M. N. 1977. *The Food Crisis in Prehistory: Overpopulation and the Origins of Agriculture.* Yale University Press, New Haven.

Coles, J. M., and E. S. Higgs. 1969. *The Archaeology of Early Man.* Faber, London.

Coltman, J. W. 1988. "The Transformer." *Scientific American* 258(1):86–95.

Committee for the Compilation of Materials on Damage Caused by the Atomic Bombs in Hiroshima and Nagasaki. 1981. *Hiroshima and Nagasaki.* Basic Books, New York.

Committee on Science, Engineering, and Public Policy. 1991. *Policy Implications of Greenhouse Warming.* National Academy Press, Washington, D.C.

Conklin, H. C. 1957. *Hanunoo Agriculture.* FAO, Rome.

Constant, E. W. 1981. *The Origins of the Turbojet Revolution.* Johns Hopkins University Press, Baltimore.

Cook, E. 1976. *Man, Energy, Society.* W. H. Freeman, San Francisco.

Corcoran, E. 1991. "Calculating Reality." *Scientific American* 262(1):100–109.

Coursen, D. L. 1992. "Explosion and Explosives." In *McGraw-Hill Encyclopedia of Science and Technology,* McGraw-Hill, New York, pp. 560–567.

Cowan, C. W., and P. J. Watson, eds. 1992. *The Origins of Agriculture.* Smithsonian Institution Press, Washington, D.C.

Cowan, R. 1990. "Nuclear Power Reactors: A Study in Technological Lock-in." *Journal of Economic History* 50:541–567.

Craig, W. 1973. *Enemy at the Gates: The Battle for Stalingrad.* Reader's Digest Press, New York.

Croil, J. 1898. *Steam Navigation.* William Briggs, Toronto.

Crossley, D. W. 1981. "Medieval Ironmaking." In D. W. Crossley, ed., *Medieval Industry,* Council for British Archaeology, London, pp. 29–41.

———. 1990. *Post-medieval Archaeology in Britain.* Leicester University Press, London.

Dalby, W. E. 1920. *Steam Power.* Edward Arnold, London.

Dartnell, J. 1978. "Coke in the Blast Furnace." *Ironmaking and Steelmaking* 5(1):18–22.

Daumas, M., ed. 1969. *A History of Technology and Invention.* Crown Publishers, New York.

David, P. A. 1991. "The Hero and the Herd in Technological History: Reflections on Thomas Edison and the Battle of the Systems." In P. Higonnet, D. S. Landes, and H.

Rosovsky, eds., *Favorites of Fortune,* Harvard University Press, Cambridge, Mass., pp. 72–119.

Davies, N. 1987. *The Aztec Empire: The Toltec Resurgence.* University of Oklahoma Press, Norman, Okla.

de Beaune, S. A., and R. White. 1993. "Ice Age Lamps." *Scientific American* 266(3):108–113.

de Camp, L. S. 1960. *The Ancient Engineers.* Dorset Press, New York.

Denevan, W. H. 1982. "Hydraulic Agriculture in the American Tropics: Forms, Measures, and Recent Research." In K. V. Flannery, ed., *Maya Subsistence,* Academic Press, New York, pp. 181–203.

Dent, A. 1974. *The Horse.* Holt, Rinehart and Winston, New York.

Derry, T. K., and T. I. Williams. 1960. *A Short History of Technology.* Oxford University Press, Oxford.

Dertouzos, M.L.D. 1991. "Communication, Computers and Networks." *Scientific American* 265(3):62–69.

Devine, W. D. 1983. "From Shafts to Wires: Historical Perspective on Electrification." *The Journal of Economic History* 63:347–372.

DeZeeuw, J. W. 1978. "Peat and the Dutch Golden Age." *AAG Bijdragen* 21:3–31.

Dickinson, H. W. 1939. *A Short History of the Steam Engine.* Cambridge University Press, Cambridge.

——. 1967. *James Watt: Craftsman and Engineer.* A. M. Kelly, New York.

Diderot, D., and J. L. D'Alembert. 1769–1772. *L'Encyclopedie ou Dictionnaire Raissonne des Sciences des Arts et des Metiers.* Avec Approbation and Privilege du Roy, Paris.

Dieffenbach, E. M., and R. B. Gray. 1960. "The Development of the Tractor." *Agricultural Yearbook* (1960):24–45.

Domros, M., and G. Peng. 1988. *The Climate of China.* Springer-Verlag, Berlin.

Doorenbos, J., and A. H. Kassam. 1979. *Yield Response to Water.* FAO, Rome.

Dowson, D. 1973. "Tribology Before Columbus." *Mechanical Engineering* 95(4):12–20.

Duby, G. 1968. *Rural Economy and Country Life in the Medieval West.* Edward Arnold, London.

——, ed. 1976. *Histoire de la France Rurale.* Seuil, Paris.

Duncan-Jones, R. 1990. *Structure and Scale in the Roman Economy.* Cambridge University Press, Cambridge.

Durand, J. D. 1960. "The Population Statistics of China, A.D. 2–1953." *Population Studies* 13:209–256.

Durnin, J.V.G.A., and R. Passmore. 1967. *Energy, Work and Leisure.* Heinemann Educational Books, London.

Earl, D. 1973. *Charcoal and Forest Management.* Oxford University Press, Oxford.

Eavenson, H. N. 1942. *The First Century and a Quarter of American Coal Industry.* Private print, Pittsbugh, Penn.

Edwards, R. D., and T. D. Williams. 1957. *The Great Famine: Studies in Irish History, 1845–1852.* New York University Press, New York.

Egg, E., J. Jobe, H. Lachouque, P. E. Cleator, and D. Reichel. 1971. *Guns.* New York Graphic Society, Greenwich, Conn.

Eisenberg, J. F. 1981. *The Mammalian Radiations.* University of Chicago Press, Chicago.

Electricity Council, the. 1973. *Electricity Supply in Great Britain.* Electricity Council, London.

Ellers, F. S. 1982. "Advanced Offshore Oil Platforms." *Scientific American* 246(4):38–49.

Ellis, C. H. 1977. *The Lore of the Train*. Crescent Books, New York.

Ellison, R. 1981. "Diet in Mesopotamia: The Evidence of the Barley Ration Texts." *Iraq* 45:35–45.

Ellul, J. 1964. *The Technological Society*. A. A. Knopf, New York.

Elton, A. 1958. "Gas for Light and Heat." In C. Singer, E. J. Holmyard, A. R. Hall, and T. I. Williams, eds., *A History of Technology*, Vol. 4, Oxford University Press, Oxford, pp. 258–275.

Engels, F. 1887. *The Condition of the Working Class in England in 1844*. Lowell Company, New York.

Esmay, M. L., and C. W. Hall, eds. 1968. *Agricultural Mechanization in Developing Countries*. Shin-Norinsha, Tokyo.

Evans, O. 1795. *The Young Mill-wright and Miller's Guide*. Published by the author, Philadelphia.

Evelyn, J. 1706. *Silva*. R. Scott, London.

Ewbank, T. 1870. *A Descriptive and Historical Account of Hydraulic and Other Machines for Raising Water*. Scribner, New York.

Falkenstein, A. 1939. *Zehnter vorläufiger Bericht über die von der Notgemeinschaft der deutschen Wissenschaft in Uruk-Warka unternommen Ausgrabungen*. Verlag der Akademie der Wissenschaften, Berlin.

Falkus, M. E. 1972. *The Industrialization of Russia, 1700–1914*. Macmillan, London.

FAO (Food and Agriculture Organization). 1980. *China: Multicropping and Related Crop Production Technology*. FAO, Rome.

_____. 1990. *Fertilizer Yearbook*. FAO, Rome.

_____. 1992. *Production Yearbook*. FAO, Rome.

FAO/WHO/UNU. 1985. *Energy and Protein Requirements*. World Health Organization (WHO), Geneva.

Farey, J. 1827. *A Treatise on the Steam Engine*. Longman, Rees, Orme, Brown and Green, London.

Federal Power Commission. 1964. *National Power Survey*. U.S. Government Printing Office, Washington, D.C.

Ferguson, E. S. 1971. "The Measurement of the 'Man-day.'" *Scientific American* 225(4):96–103.

Feynman, R. 1988. *The Feynman Lectures on Physics*. Addison-Wesley, Reading, Mass.

Finley, M. I. 1959. "Was Greek Civilisation Based on Slave Labor?" *Historia* 8:145–164.

_____. 1965. "Technical Innovation and Economic Progress in the Ancient World." *Economic History Review* 18:29–45.

Finniston, M., T. Williams, and C. Bissell, eds. 1992. *Oxford Illustrated Encyclopedia of Invention and Technology*. Oxford University Press, Oxford.

Fitchen, J. 1961. *The Construction of Gothic Cathedrals*. University of Chicago Press, Chicago.

Flannery, K. V., ed. 1982. *Maya Subsistence*. Academic Press, New York.

Flink, J. J. 1988. *The Automobile Age*. MIT Press, Cambridge, Mass.

Flower, R., and M. W. Jones. 1981. *One Hundred Years of Motoring*. McGraw-Hill, Maidenhead.

Fluck, R. C., ed. 1992a. *Energy in Farm Production*. Elsevier, Amsterdam.

————. 1992b. "Energy of Human Labor." In R. C. Fluck, ed., *Energy in Farm Production,* Elsevier, Amsterdam, pp. 31–37.

Foley, R. A., and C. Lee. 1992. "Ecology and Energetics of Encephalization in Hominid Evolution." In A. Whiten and E. M. Widdowson, eds., *Foraging Strategies and Natural Diet of Monkeys, Apes and Humans,* Clarendon Press, Oxford, pp. 63–71.

Forbes, R. J. 1956. "Metallurgy." In C. Singer, E. J. Holmyard, A. R. Hall, and T. I. Williams, eds., *A History of Technology,* Vol. 2, Oxford University Press, Oxford, pp. 41–80.

————. 1958. "Power to 1850." In C. Singer, E. J. Holmyard, A. R. Hall, and T. I. Williams, eds., *A History of Technology,* Vol. 4, Oxford University Press, Oxford, pp. 148–167.

————. 1964–1972. *Studies in Ancient Technology.* E. J. Brill, Leiden.

————. 1964. "Bitumen and Petroleum in Antiquity." In *Studies in Ancient Technology,* Vol. 1. E. J. Brill, Leiden, pp. 1–124.

————. 1965. *Studies in Ancient Technology,* Vol. 2. E. J. Brill, Leiden.

————. 1966. "Heat and Heating." In *Studies in Ancient Technology,* Vol. 6. E. J. Brill, Leiden, pp. 1–103.

————. 1972. "Copper." In *Studies in Ancient Technology,* Vol. 9. E. J. Brill, Leiden, pp. 1–133.

Fortune. 1993. *The Fortune 500.* Fortune, New York.

Fox, R. E. 1988. *Energy and the Evolution of Life.* W. H. Freeman, New York.

Fox, W. 1976. *The Mill.* McClelland and Stewart, Toronto.

Fraenkel, P. L. 1986. *Water Lifting Devices.* FAO, Rome.

Francis, D. 1990. *The Great Chase: A History of World Whaling.* Penguin Books, Toronto.

Freeman, C., ed. 1983. *Long Waves in the World Economy.* Frances Pinter, London.

Freeman, L. G. 1981. "The Fat of the Land: Notes on Paleolithic Diet in Iberia." In R.S.O. Harding and G. Teleki, eds., *Omnivorous Primates,* Columbia University Press, New York, pp. 104–165.

Freese, S. 1957. *Windmills and Millwrighting.* Cambridge University Press, Cambridge.

Friedman, H. B. 1992. "DDT (Dichlorodiphenyltrichloroethane)." *Journal of Chemical Education* 69:362–365.

Frison, G. C. 1987. "Prehistoric Hunting Strategies." In M. H. Nitecki and D. V. Nitecki, eds., *The Evolution of Human Hunting,* Plenum Press, New York, pp. 177–223.

Fry, H. 1896. *The History of North Atlantic Steam Navigation.* Sampson, Low, Marston & Co., London.

Fussell, G. E. 1972. *The Classical Tradition in West European Farming.* Fairleigh Dickinson University Press, Rutherford.

Gamarra, N. T. 1969. *Erroneous Predictions and Negative Comments ...* Legislative Reference Service, The Library of Congress, Washington, D.C.

Gardner, J., and J. Maier. 1984. *Gilgamesh.* A. A. Knopf, New York.

Geertz, C. 1963. *Agricultural Involution.* University of California Press, Berkeley.

Georgescu-Roegen, N. 1980. "Afterword." In J. Rifkin, *Entropy,* Viking Press, New York, pp. 261–269.

Gille, B., ed. 1978. *Histoire des Techniques.* Gallimard, Paris.

Gold, B., W. S. Pierce, G. Rosegger, and M. Perlman. 1984. *Technological Progress and Industrial Leadership: The Growth of the U.S. Steel Industry, 1900–1970.* Lexington Books, Lexington, Mass.

Goldsmith, R. W. 1946. "The Power of Victory: Munitions Output in World War II." *Military Affairs* 10:69–80.

Goldstein, J. 1985. "Basic Human Needs: The Plateau Curve." *World Development* 13:595–609.

Goudsblom, J. 1992. *Fire and Civilization*. Allen Lane, London.

Greenwood, W. H. 1907. *Iron*. Cassell and Company, London.

Grigg, D. B. 1974. *The Agricultural Systems of the World*. Cambridge University Press, Cambridge.

Grimal, N. 1992. *A History of Ancient Egypt*. Blackwell, Oxford.

Grousset, R. 1970. *The Empire of the Steppes*. Rutgers University Press, New Brunswick, N.J.

Grzimek, B., ed. 1972. *Grzimek's Animal Life Encyclopedia*. Van Nostrand Reinhold, New York.

Gunston, B. 1986. *World Encyclopaedia of Aero Engines*. Patrick Stephens, Wellingborough.

Haaland, R., and P. Shinnie, eds. 1985. *African Iron Working—Ancient and Traditional*. Oxford University Press, Oxford.

Hadfield, C. 1969. *The Canal Age*. Praeger Publishers, New York.

Hadingham, E. 1992. "Pyramid Schemes." *The Atlantic* 270(5):38–42, 51–52.

Hair, T. H. 1844. *Sketches of the Coal Mines in Northumberland and Durham*. J. Madden & Company, London.

Hall, D. O., G. W. Barnard, and P. A. Moss. 1982. *Biomass for Energy in the Developing Countries*. Pergamon Press, Oxford.

Hammond, N. 1986. "The Emergence of Maya Civilization." *Scientific American* 255(2):106–115.

Haney, L. H. 1968. *A Congressional History of Railways in the United States*. A. M. Kelley, New York.

Hanna, J. M., and D. E. Brown. 1983. "Human Heat Tolerance: An Anthropological Perspective." *Annual Review of Anthropology* 12:259–284.

Harako, R. 1981. "The Cultural Ecology of Hunting Behavior Among Mbuti Pygmies of the Ituri Forest, Zaire." In R.S.O. Harding and G. Teleki, eds., *Omnivorous Primates*, Columbia University Press, New York, pp. 499–555.

Harris, J. R. 1974. "The Rise of Coal Technology." *Scientific American* 233(2):92–97.

———. 1988. *The British Iron Industry 1700–1850*. Macmillan, London.

Harris, M. 1966. "The Cultural Ecology of India's Sacred Cattle." *Current Anthropology* 7:51–66.

Harrison, M. 1988. "Resource Mobilization for World War II: The U.S.A., U.K., U.S.S.R., and Germany, 1938–1945." *Economic History Review* 41:171–192.

Harrison, P. D., and B. L. Turner II, eds. 1978. *Pre-Hispanic Maya Agriculture*. University of New Mexico Press, Albuquerque.

Hassan, F. A. 1984. "Environment and Subsistence in Predynastic Egypt." In J. D. Clark and S. A. Brandt, eds., *From Hunters to Farmers: The Causes and Consequences of Food Production in Africa*, University of California Press, Berkeley, Calif., pp. 57–64.

Haudricourt, A. G., and M.J.B. Delamarre. 1955. *L'Homme et la Charrue a travers le Monde*. Gallimard, Paris.

Hayden, B. 1981. "Subsistence and Ecological Adaptations of Modern Hunter/Gatherers." In R.S.O. Harding and G. Teleki, eds., *Omnivorous Primates*, Columbia University Press, New York, pp. 344–421.

Headland, T. N., and L. A. Reid. 1989. "Hunter-Gatherers and Their Neighbors from Pre-history to the Present." *Current Anthropology* 30:43–66.

Heal, D. W. 1975. "Modern Perspectives on the History of Fuel Economy in the Iron and Steel Industry." *Ironmaking and Steelmaking* 2(4):222–227.

Heidenreich, C. 1971. *Huronia: A History and Geography of Huron Indians, 1600–1650.* McClelland and Stewart, Toronto.

Heizer, R. F. 1966. "Ancient Heavy Transport, Methods and Achievements." *Science* 153:821–830.

Helland, J. 1980. *Five Essays on the Study of Pastoralists and the Development of Pastoralism.* Universitet i Bergen, Bergen.

Helsel, Z., ed. 1987. *Energy in Plant Nutrition and Pest Control.* Elsevier, Amsterdam.

Heston, A. 1971. "An Approach to the Sacred Cow of India." *Current Anthropology* 12:191–209.

Hill, D. 1984. *A History of Engineering in Classical and Medieval Times.* Open Court Publishing, La Salle, Ill.

Hindle, B., ed. 1975. *America's Wooden Age: Aspects of Its Early Technology.* Sleepy Hollow Restorations, Tarrytown, N.Y.

Hitchcock, R. K., and J. I. Ebert. 1984. "Foraging and Food Production Among Kalahari Hunter/Gatherers." In J. D. Clark and S. A. Brandt, eds., *From Hunters to Farmers: The Causes and Consequences of Food Production in Africa,* University of California Press, Berkeley, Calif., pp. 328–348.

Ho, Ping-ti. 1975. *The Cradle of the East.* The Chinese University of Hong Kong Press, Hong Kong.

Hockett, C. F., and R. Ascher. 1964. "The Human Revolution." *Current Anthropology* 5:135–168.

Hodge, A. T. 1985. "Siphons in Roman Aqueducts." *Scientific American* 252(6):114–119.

———. 1992. *Roman Aqueducts and Water Supply.* Duckworth, London.

Hodges, H. 1990. *Technology in the Ancient World.* Barnes and Noble, New York.

Hodges, P. 1989. *How the Pyramids Were Built.* Element Books, Rockport, Mass.

Hogan, W. T. 1971. *Economic History of the Iron and Steel Industry in the United States.* Lexington Books, Lexington, Mass.

Holley, I. B. 1964. *Buying Aircraft: Materiel Procurement for the Army Air Forces.* Department of the Army, Washington, D.C.

Hommel, R. P. 1937. *China at Work.* The Bucks County Historical Society, Doylestown, Penn.

Hopfen, H. J. 1969. *Farm Implements for Arid and Tropical Regions.* FAO, Rome.

Houghton, J. T., G. J. Jenkins, and J. J. Ephraums, eds. 1990. *Climate Change: The IPCC Scientific Assessment.* Cambridge University Press, Cambridge.

Hounshell, D. A. 1981. "Two Paths to the Telephone." *Scientific American* 244(1):156–164.

Howell, J. M. 1987. "Early Farming in Northwestern Europe." *Scientific American* 257(5):118–126.

Hua, Jue-ming. 1983. "The Mass Production of Iron Castings in Ancient China." *Scientific American* 258(1):120–128.

Hughes, T. P. 1983. *Networks of Power.* Johns Hopkins University Press, Baltimore.

Hunter, L. C. 1975. "Water Power in the Century of Steam." In B. Hindle, ed., *America's Wooden Age: Aspects of Its Early Technology,* Sleepy Hollow Restorations, Tarrytown, N.Y., pp. 160–192.

Hyde, C. K. 1977. *Technological Change and the British Iron Industry, 1700–1870.* Princeton University Press, Princeton, N.J.

Hyland, A. 1990. *Equus: The Horse in the Roman World.* Yale University Press, New Haven.

Jakab, P. L. 1990. *Visions of a Flying Machine: The Wright Brothers and the Process of Invention.* Smithsonian Institution Press, Washington, D.C.

James, S. R. 1989. "Hominid Use of Fire in the Lower and Middle Pleistocene." *Current Anthropology* 30:1–26.

James, T.G.H. 1984. *Pharaoh's People.* University of Chicago Press, Chicago.

Jarman, M. R., G. N. Bailey, and H. N. Jarman, eds. 1982. *Early European Agriculture.* Cambridge University Press, Cambridge.

Jehl, F. 1937. *Menlo Park Reminiscences.* Edison Institute, Dearborn, Mich.

Jensen, H. 1969. *Sign, Symbol and Script.* G. P. Putnam's Sons, New York.

Jochim, M. A. 1976. *Hunter-Gatherer Subsistence and Settlement: A Predictive Model.* Academic Press, New York.

Johannsen, O. 1953. *Geschichte des Eisens.* Stahleisen, Düsseldorf.

Johnson, A. W., and T. Earle. 1987. *The Evolution of Human Societies.* Stanford University Press, Stanford, Calif.

Johnson, E. D. 1973. *Communication.* Scarecrow Press, Metuchen, N.J.

Jones, H. 1973. *Steam Engines.* Ernest Benn, London.

Jones, H. M. 1971. *The Age of Energy.* Viking Press, New York.

Jope, E. M. 1956. "Vehicles and Harness." In C. Singer, E. J. Holmyard, A. R. Hall, and T. I. Williams, eds., *A History of Technology,* Vol. 2, Oxford University Press, Oxford, pp. 537–562.

Josephson, M. 1959. *Edison: A Biography.* McGraw-Hill, New York.

Kay, J. P. 1832. *The Moral and Physical Conditions of the Working Classes Employed in the Cotton Manufacture in Manchester.* J. Ridgway, London.

Kelly, R. L. 1983. "Hunter-Gatherer Mobility Strategies." *Journal of Anthropological Research* 39:277–306.

Kendall, A. 1973. *Everyday Life of the Incas.* B. T. Batsford, London.

Kesaris, P., ed. 1977. *Manhattan Project: Official History and Documents.* University Publications of America, Washington, D.C.

Khazanov, A. M. 1984. *Nomads and the Outside World.* Cambridge University Press, Cambridge.

King, C. D. 1948. *Seventy-five Years of Progress in Iron and Steel.* American Institute of Mining and Metallurgical Engineers, New York.

King, F. H. 1927. *Farmers of Forty Centuries.* Harcourt, Brace & Co., New York.

Kirk, R. E., and D. F. Othmer, eds. 1947. *Encyclopedia of Chemical Technology.* The Interscience Encyclopedia, New York.

Kirk, R. L. 1981. *Aboriginal Man Adapting.* Clarendon Press, Oxford.

Klein, H. A. 1978. "Pieter Bruegel the Elder as a Guide to 16th Century Technology." *Scientific American* 238(3):134–140.

Klemm, F. 1964. *A History of Western Technology.* MIT Press, Cambridge, Mass.

Klima, B. 1954. "Paleolithic Huts at Dolni Vestonice, Czechoslovakia." *Antiquity* 28:4–14.

Kloss, E. 1963. *Der Luftkrieg über Deutschland, 1939–1945.* DTV, Munchen.

Komlos, J. 1988. "Agricultural Productivity in America and Eastern Europe: A Comment." *The Journal of Economic History* 48:665–664.

Krenz, J. H. 1976. *Energy Conversion and Utilization*. Allyn and Bacon, Boston.

Kuznets, S. 1971. *Economic Growth of Nations: Total Output and Production Structure*. Harvard University Press, Cambridge, Mass.

Lacey, J. M. 1935. *A Comprehensive Treatise on Practical Mechanics*. Technical Press, London.

Landels, J. G. 1980. *Engineering in the Ancient World*. Chatto & Windus, London.

Landes, D. S. 1961. *The Unbound Prometheus*. Cambridge University Press, Cambridge.

Lane, F. C. 1934. *Venetian Ships and Shipbuilders of the Renaissance*. Johns Hopkins University Press, Baltimore.

Langdon, J. 1986. *Horses, Oxen and Technological Innovation*. Cambridge University Press, Cambridge.

Leach, E. R. 1959. "Hydraulic Society in Ceylon." *Past & Present* 15:2–26.

Leakey, R., and R. Lewin. 1992. *Origins Reconsidered*. Doubleday, New York.

Legge, A. J., and P. A. Rowley-Conwy. 1987. "Gazelle Killing in Stone Age Syria." *Scientific American* 257(2):88–95.

Lepre, J. P. 1990. *The Egyptian Pyramids*. McFarland & Company, Jefferson, N.C.

Leser, P. 1931. *Entstehung und Verbreitung des Pfluges*. Aschendorff, Münster.

Lesser, I. O. 1991. *Oil, the Persian Gulf, and Grand Strategy*. Rand Corporation, Santa Monica, Calif.

Liddell Hart, B.H.L. 1954. *Strategy*. Faber & Faber, London.

———. 1970. *History of the Second World War*. G. P. Putnam's Sons, New York.

Lilienthal, D. E. 1944. *TVA—Democracy on the March*. Harper & Brothers, New York.

Lindsay, J. 1974. *Blast-power and Ballistics*. Harper & Row, New York.

Lindsay, R. B. 1975. *Energy: Historical Development of the Concept*. Dowden, Hutchinson & Ross, Stroudsburg, Penn.

Ling, P. J. 1990. *America and the Automobile: Technology, Reform and Social Change*. Manchester University Press, Manchester.

Livi-Bacci, M. 1991. *Population and Nutrition*. Cambridge University Press, Cambridge.

Lizerand, G. 1942. *Le regime rural de l'ancienne France*. Presses Universitaires, Paris.

Lizot, J. 1977. "Population, Resources and Warfare Among the Yanomami." *Man* 12:497–517.

Lotka, A. 1925. *Elements of Physical Biology*. Williams and Wilkins, Baltimore.

Lowrance, R., B. R. Stinner, and G. J. House, eds. 1984. *Agricultural Ecosystems*. Wiley, New York.

L'vovich, M. I., and G. F. White. 1990. "Use and Transformation of Terrestrial Water Systems." In B. L. Turner II, W. C. Clark, R. W. Kates, J. F. Richards, J. T. Matthews, and W. B. Meyer, eds., *The Earth as Transformed by Human Action*, Cambridge University Press, Cambridge, pp. 235–252.

Lyman, T., ed. 1961. *Metals Handbook*. American Society for Metals, Novelty, Ohio.

Maddin, R., ed. 1988. *The Beginning of the Use of Metals and Alloys*. MIT Press, Cambridge, Mass.

Marchetti, C. 1986. "Fifty-year Pulsation in Human Affairs." *Futures* 18:376–388.

Marcus, J. 1982. "The Plant World of the Sixteenth- and Seventeenth-century Lowland Maya." In K. V. Flannery, ed., *Maya Subsistence*, Academic Press, New York, pp. 239–273.

Mark, J. 1985. "Changes in the British Brewing Industry in the Twentieth Century." In D. J. Oddy and D. P. Miller, eds., *Diet and Health in Modern Britain*, Croom Helm, London, pp. 81–101.

Mayr, O. 1970. *The Origins of Feedback Control.* MIT Press, Cambridge, Mass.

McCarthy, F. D. 1967. *Australian Aboriginal Stone Implements.* Australian Museum, Sydney.

McCloy, S. T. 1952. *French Inventions of the Eighteenth Century.* University of Kentucky Press, Lexington, Ky.

McCormick, C. 1931. *The Century of the Reaper.* Houghton Mifflin, Boston.

McEvedy, C., and R. Jones. 1978. *Atlas of World Population History.* Allen Lane, London.

McKeown, T. 1976. *The Modern Rise of Population.* Arnold, London.

McNeill, W. H. 1980. *The Human Condition.* Princeton University Press, Princeton, N.J.

———. 1989. *The Age of Gunpowder Empires, 1450–1800.* American Historical Association, Washington, D.C.

———. 1993. Personal communication.

Meindl, J. D. 1987. "Chips for Advanced Computers." *Scientific American* 257(4):78–88.

Melko, M. 1971. "Is Western Civilization Unique?" *Civilisations* 21:38–43.

Mellars, P. A. 1985. "The Ecological Basis of Social Complexity in the Upper Paleolithic of Southwestern France." In T. D. Price and J. A. Brown, eds., *Prehistoric Hunter-Gatherers,* Academic Press, Orlando, Fla., pp. 271–297.

Melosi, M. V. 1982. "Energy Transitions in the Nineteenth-century Economy." In G. H. Daniels and M. H. Rose, eds., *Energy and Transport,* Sage Publications, Beverly Hills, Calif., pp. 55–67.

Melville, H. 1851. *Moby-Dick or the Whale.* Harper & Brothers, New York.

Mendels, F. 1972. "Protoindustrialization: The First Phase of Industrialization Process." *Journal of Economic History* 32:241–261.

Mendelssohn, K. 1974. *The Riddle of the Pyramids.* Thames and Hudson, London.

Mensch, G. 1979. *Stalemate in Technology.* Ballinger, Cambridge, Mass.

Minchinton, W. 1980. "Wind Power." *History Today* 30(3):31–36.

Minchinton, W., and P. Meigs. 1980. "Power from the Sea." *History Today* 30(3):42–46.

Mitchell, B. R. 1975. *European Historical Statistics, 1750–1970.* Columbia University Press, New York.

Mitchell, W. A. 1931. *Outlines of the World's Military History.* Military Service Publishing, Harrisburg, Penn.

Mokyr, J. 1976. *Industrialization in the Low Countries, 1795–1850.* Yale University Press, New Haven.

Molenaar, A. 1956. *Water Lifting Devices for Irrigation.* FAO, Rome.

Montet, P. 1966. *Egipto Eterno.* Ediciones Guadarrama, Madrid.

Moreau, R. 1984. *The Computer Comes of Age.* MIT Press, Cambridge, Mass.

Moritz, L. A. 1958. *Grain-Mills and Flour in Classical Antiquity.* Oxford University Press, Oxford.

Morrison, J. S., and J. F. Coates. 1986. *The Athenian Trireme.* Cambridge University Press, Cambridge.

Morrison, J. S., and R. T. Williams. 1968. *Greek Oared Ships, 900–322 B.C.* Cambridge University Press, Cambridge.

Mukerji, C. 1981. *From Graven Images: Pattern of Modern Materialism.* Columbia University Press, New York.

Mumford, L. 1934. *Technics and Civilization.* Harcourt, Brace and Company, New York.

Murdock, G. P. 1967. "Ethnographic Atlas." *Ethnology* 6:109–236.

Murra, J. V. 1980. *The Economic Organization of the Inka State.* JAI Press, Greenwich, Conn.

MVMA (Motor Vehicle Manufacturing Association). 1992. *Facts & Figures '92*. MVMA, Detroit, Mich.

National Research Council. 1986. *Electricity in Economic Growth*. National Academy Press, Washington, D.C.

Naville, E. 1908. *The Temple of Deir el Bahari*, Part VI, The Egyptian Exploration Fund, London.

Needham, J. 1964. *The Development of Iron and Steel Technology in China*. W. Heffer & Sons, Cambridge.

Needham, J., et al. 1954– . *Science and Civilization in China*. Cambridge University Press, Cambridge.

————. 1965. *Science and Civilisation in China. Volume 4, Part II: Physics and Physical Technology*. Cambridge University Press, Cambridge.

————. 1971. *Science and Civilisation in China. Volume 4, Part III: Civil Engineering and Nautics*. Cambridge University Press, Cambridge.

————. 1986. *Science and Civilisation in China. Volume 5, Part VII: Military Technology: The Gunpowder Epic*. Cambridge University Press, Cambridge.

Nef, J. U. 1932. *The Rise of the British Coal Industry*. George Routledge & Sons, London.

Neuburger, A. 1930. *The Technical Arts and Sciences of the Ancients*. Barnes and Noble, New York.

Niel, F. 1961. *Dolmens et Menhirs*. Presses Universitaires de France, Paris.

des Noettes, R.J.E.C.L. 1931. *L'Attelage et le Cheval de Selle a travers les Ages*. Picard, Paris.

Nye, D. E. 1992. *Electrifying America*. MIT Press, Cambridge, Mass.

O'Brien, P., ed. 1983. *Railways and the Economic Development of Western Europe, 1830–1914*. St. Martin's Press, New York.

O'Connell, J. F., and K. Hawkes. 1984. "Food Choice and Foraging Sites Among the Alyawara." *Journal of Anthropological Research* 40:504–535.

Odend'hal, S. 1972. "Energetics of Indian Cattle in Their Environment." *Human Ecology* 1:3–22.

Odum, H. T. 1971. *Environment, Power, and Society*. John Wiley, New York.

OECD. 1970–. *Economic Indicators*. OECD, Paris.

Oleson, J. P. 1984. *Greek and Roman Mechanical Water-Lifting Devices: The History of a Technology*. University of Toronto Press, Toronto.

Olson, M. 1982. *The Rise and Fall of Nations*. Yale University Press, New Haven.

Openshaw, K. 1978. "Woodfuel—A Time for Re-assessment." *Natural Resources Forum* 3:35–51.

Orme, B. 1977. "The Advantages of Agriculture." In J.V.S. Megaw, ed., *Hunters, Gatherers and First Farmers Beyond Europe*, Leicester University Press, Leicester, pp. 41–49.

Ostwald, W. 1909. *Energetische Grundlagen der Kulturwissenschaft*. Alfred Kroner, Leipzig.

Pacey, A. 1990. *Technology in World Civilization*. MIT Press, Cambridge, Mass.

Patwardhan, S. 1973. *Change Among India's Harijans*. Orient Longman, New Delhi.

Paul, C. H. 1983. *Irrigation*. The Irrigation Association, Arlington, Va.

Peacey, J. G. 1979. *The Iron Blast Furnace*. Pergamon Press, Oxford.

Pentzer, W. T. 1966. "The Giant Job of Refrigeration." In *USDA Yearbook*, U.S. Department of Agriculture, Washington, D.C, pp. 123–138.

Perdue, P. C. 1987. *Exhausting the Earth*. Harvard University Press, Cambridge, Mass.

Perkins, D. H. 1969. *Agricultural Development in China, 1368–1968*. Aldine, Chicago.

Perrodon, A. 1985. *Histoire des Grandes Decouvertes Petrolieres.* Elf Aquitaine, Paris.

Petroski, H. 1993. "On Dating Inventions." *American Scientist* 81:314–318.

Piggott, S. 1983. *The Earliest Wheeled Transport.* Cornell University Press, Ithaca, N.Y.

Pimentel, D., ed. 1980. *Handbook of Energy Utilization in Agriculture.* CRC Press, Boca Raton, Fla.

Pokotylo, D. L. 1988. *Blood from Stone.* UBC Museum of Anthropology, Vancouver.

Pope, S. T. 1923. "A Study of Bows and Arrows." *University of California Publications in American Archaeology and Ethnology* 13:329–414.

Porada, E. 1965. *The Art of Ancient Iran.* Crown, New York.

Priamo, C. 1976. *Mills of Canada.* McGraw-Hill Ryerson, Toronto.

Price, T. D. 1991. "The Mesolithic of Northern Europe." *Annual Review of Anthropology* 20:211–233.

Price, T. D., and J. A. Brown, eds. 1985. *Prehistoric Hunter-Gatherers.* Academic Press, Orlando, Fla.

Prost, A. 1991. "Public and Private Spheres in France." In A. Prost and G. Vincent, eds., *A History of Private Life. Volume 5: Riddles of Identity in Modern Times,* Harvard University Press, Cambridge, Mass., pp. 1–143.

Protzen, J-P. 1986. "Inca Stonemasonry." *Scientific American* 254(2):94–105.

Pryor, F. L. 1983. "Causal Theories About the Origin of Agriculture." *Research in Economic History* 8:93–124.

Quill, H. 1966. *John Harrison: The Man who Found Longitude.* Baker, London.

Ralston, A., ed. 1976. *Encyclopedia of Computer Science.* Van Nostrand Reinhold, New York.

Ramelli, A. 1588. *Le Diverse et Artificiose Machine.* Translated by M. Teach Gnudi. Johns Hopkins University Press, Baltimore (1976).

Rankine, W.J.M. 1866. *Useful Rules and Tables Relating to Mensuration, Engineering Structures and Machines.* G. Griffin & Company, London.

Rappaport, R. A. 1968. *Pigs for the Ancestors.* Yale University Press, New Haven.

Ratcliffe, K. 1985. *Liquid Gold Ships: History of the Tanker (1859–1984).* Lloyds, London.

Reed, C. A., ed. 1977. *Origins of Agriculture.* Mouton, The Hague.

Revel, J. 1979. "A Capital City's Privileges: Food Supplies in Early-modern Rome." In R. Forster and O. Ranum, eds., *Food and Drink in History,* Johns Hopkins University Press, Baltimore, pp. 37–49.

Reynolds, J. 1970. *Windmills and Watermills.* Hugh Evelyn, London.

Richards, J. F. 1990. "Land Transformation." In B. L. Turner II, W. C. Clark, R. W. Kates, J. F. Richards, J. T. Matthews, and W. B. Meyer, eds., *The Earth as Transformed by Human Action,* Cambridge University Press, New York, pp. 163–178.

Rindos, D. 1984. *The Origin of Agriculture: An Ecological Perspective.* Academic Press, Orlando, Fla.

Robinson, E., and A. E. Musson. 1969. *James Watt and the Steam Revolution.* Augustus M. Kelley, New York.

Rogin, L. 1931. *The Introduction of Farm Machinery.* University of California Press, Berkeley.

Rollins, A. M. 1983. *The Fall of Rome.* McFarland & Co., Jefferson, N.C.

Romer, A. S. 1959. *The Vertebrate Story.* University of Chicago Press, Chicago.

Rosenberg, N. 1975. "America's Rise to Woodworking Leadership." In B. Hindle, ed., *America's Wooden Age: Aspects of Its Early Technology,* Sleepy Hollow Restorations, Tarrytown, N.Y., pp. 37–62.

Rosenberg, N., and L. E. Birdzell, Jr. 1986. *How the West Grew Rich: The Economic Transformation of the Industrial World.* Basic Books, New York.

Ross, S. 1989. *Soil Processes: A Systematic Approach.* Routledge, London.

Rostow, W. W. 1965. *The Stages of Economic Growth.* Cambridge University Press, Cambridge.

———. 1978. *The World Economy.* University of Texas Press, Austin.

Rouse, J. E. 1970. *World Cattle.* University of Oklahoma Press, Norman.

Rowland, F. S. 1989. "Chlorofluorocarbons and the Depletion of Stratospheric Ozone." *American Scientists* 77:36–45.

Ruddle, K., and G. Zhong. 1988. *Integrated Agriculture-Aquaculture in South China.* Cambridge University Press, Cambridge.

Rühlmann, G. 1962. *Kleine Geschichte der Pyramiden.* Verlag der Kunst, Dresden.

Ryder, H. W., H. J. Carr, and P. Herget. 1976. "Future Performance in Footracing." *Scientific American* 234(6):109–119.

Sagan, S. D. 1988. "Origins of the Pacific War." *Journal of Interdisciplinary History* 18:893–922.

Sahlins, M. 1972. *Stone Age Economics.* Aldine, Chicago.

Sanders, W. T., J. R. Parsons, and R. S. Santley, eds. 1979. *The Basin of Mexico.* Academic Press, New York.

Savage, C. I. 1959. *An Economic History of Transport.* Hutchinson, London.

Saxena, K. K. 1962. *Indian Railways.* Vora & Company, Bombay.

Schlebecker, J. T. 1975. *Whereby We Thrive.* Iowa State University Press, Ames.

Schmidt, P., and D. H. Avery. 1978. "Complex Iron Smelting and Prehistoric Culture in Tanzania." *Science* 201:1085–1089.

Schubert, H. R. 1958. "Extraction and Production of Metals: Iron and Steel." In C. Singer, E. J. Holmyard, A. R. Hall, and T. I. Williams, eds., *A History of Technology,* Oxford University Press, Oxford, Vol. 4, pp. 99–117, Vol. 5, pp. 53–71.

Schumpeter, J. A. 1939. *Business Cycles: A Theoretical and Statistical Analysis of the Capitalist Processes.* McGraw-Hill, New York.

Schurr, S. H. 1984. "Energy Use, Technological Change, and Productive Efficiency: An Economic-Historical Interpretation." *Annual Review of Energy* 9:409–425.

Schurr, S. H., and B. C. Netschert. 1960. *Energy in the American Economy, 1850–1975.* Johns Hopkins University Press, Baltimore.

Seaborg, G. T. 1972. "Opening Address." In *Peaceful Uses of Atomic Energy. Volume 1: United Nations,* New York, pp. 29–35.

Seavoy, R. E. 1986. *Famine in Peasant Societies.* Greenwood Press, New York.

Seebohm, M. E. 1927. *The Evolution of the English Farm.* Allen & Unwin, London.

Sellin, H. J. 1983. "The Large Roman Water Mill at Barbegal (France)." *History of Technology* 8:91–109.

Senancour, E. P. 1804. *Obermann.* Cerioux, Paris.

Sexton, A. H. 1897. *Fuel and Refractory Materials.* Blackie & Son, London.

Sheehan, G. W. 1985. "Whaling as an Organizing Focus in Northwestern Eskimo Society." In T. D. Price and J. A. Brown, eds., *Prehistoric Hunter-Gatherers,* Academic Press, Orlando, Fla., pp. 123–154.

Shen, T. H. 1951. *Agricultural Resources of China.* Cornell University Press, Ithaca, N.Y.

Shimada, I., and J. F. Merkel. 1991. "Copper-alloy Metallurgy in Ancient Peru." *Scientific American* 265(1):80–86.

Silver, C. 1976. *Guide to the Horses of the World.* Elsevier Phaidon, Oxford.

Singer, C., E. J. Holmyard, A. R. Hall, and T. I. Williams, eds. 1954–1958. *A History of Technology.* Oxford University Press, Oxford.

Singer, J. D., and M. Small. 1972. *The Wages of War 1816–1965: A Statistical Handbook.* John Wiley, New York.

Sitwell, N.H.H. 1981. *Roman Roads of Europe.* St. Martin's Press, New York.

Siuru, B. 1989. "Horsepower to the People." *Mechanical Engineering* 111(2):42–46.

Skilton, C. P. 1947. *British Windmills and Watermills.* Collins, London.

Slicher van Bath, B. H. 1963. *The Agrarian History of Western Europe, A.D. 500–1850.* Arnold, London.

Smeaton, J. 1760. *An Experimental Enquiry Concerning the Natural Powers of Water and Wind to Turn Mills, and Other Machines, Depending on a Circular Motion.* J. Smeaton, London.

Smil, V. 1981. "China's Food." *Food Policy* 6:67–77.

———. 1983. *Biomass Energies.* Plenum Press, New York.

———. 1985. *Carbon Nitrogen Sulfur: Human Interference in Grand Biospheric Cycles.* Plenum Press, New York.

———. 1986. "Food Production and Quality of Diet in China." *Population and Development Review* 12:25–45.

———. 1988. *Energy in China's Modernization.* M. E. Sharpe, New York.

———. 1991. *General Energetics.* John Wiley, New York.

———. 1992. "Agricultural Energy Costs: National Analyses." In R. C. Fluck, ed., *Energy in Farm Production,* Elsevier, Amsterdam, pp. 85–100.

———. 1993. *Global Ecology.* Routledge, London.

Smith, B. 1964. *Japan: A History in Art.* Doubleday, New York.

Smith, N. 1978. "Roman Hydraulic Technology." *Scientific American* 238(5):134–161.

———. 1980. "The Origins of the Water Turbine." *Scientific American* 242(1):138–148.

Smoluchowski, R. 1983. *The Solar System.* Scientific American Library, New York.

Smythe, R. H. 1967. *Structure of the Horse.* J. A. Allen & Co., London.

Snyder, H. 1930. *Bread.* Macmillan, New York.

Soedel, W., and V. Foley. 1979. "Ancient Catapults." *Scientific American* 240(3):150–160.

Speer, A. 1970. *Inside the Third Reich.* Macmillan, London.

Spring Rice, M. 1939. *Working-Class Wives: Their Health and Conditions.* Penguin, Harmondsworth.

Stanhill, G. 1976. "Trends and Deviations in the Yield of the English Wheat Crop During the Last 750 Years." *Agro-Ecosystems* 3:1–10.

Starbuck, A. 1878. *History of the American Whale Fishery.* A. Starbuck, Waltham, Mass.

State Statistical Bureau. 1991. *China's Statistical Yearbook 1990.* SSB, Beijing.

Steel, R., and A. P. Harvey, eds. 1979. *The Encyclopedia of Prehistoric Life.* McGraw-Hill, New York.

Stockhuyzen, F. 1963. *The Dutch Windmill.* Universe Books, New York.

Straker, E. 1969. *Wealden Iron.* A. M. Kelley, New York.

Sugawara, T., M. Ikeda, T. Shimotsuma, M. Higuchi, M. Izuka, and K. Kuroda. 1976. "Construction and Operation of No. 5 Blast Furnace, Fukuyama Works, Nippon Kokan KK." *Ironmaking and Steelmaking* 3(5):241–251.

Summers, R., and A. Heston. 1991. "The Penn World Table (Mark 5): An Expanded Set of International Comparisons, 1950–1988." *Quarterly Journal of Economics* 106:327–368.

Swade, D. 1991. *Charles Babbage and his Calculating Engines.* Science Museum, London.

Sykes, L. R., and D. M. Davis. 1987. "The Yields of Soviet Strategic Weapons." *Scientific American* 256(1):29–37.

Taeuber, I. B. 1958. *The Population of Japan.* Princeton University Press, Princeton, N.J.

Tainter, J. A. 1989. *The Collapse of Complex Societies.* Cambridge University Press, Cambridge.

Tanaka, J. 1980. *The San Hunter-Gatherers of the Kalahari.* University of Tokyo Press, Tokyo.

Taylor, C. F. 1984. *The Internal-Combustion Engine in Theory and Practice.* MIT Press, Cambridge, Mass.

Taylor, F. S. 1972. *A History of Industrial Chemistry.* Arno Press, New York.

Taylor, F. W. 1911. *Principles of Scientific Management.* Harper & Brothers, New York.

Taylor, G. R., ed. 1982. *The Inventions that Changed the World.* Reader's Digest Association, London.

Taylor, M.J.H., ed. 1989. *Jane's Encyclopedia of Aviation.* Portland House, New York.

Telleen, M. 1977. *The Draft Horse Primer.* Rodale Press, Emmaus, Penn.

Temin, P. 1964. *Iron and Steel in Nineteenth-Century America.* MIT Press, Cambridge, Mass.

Temple, R. 1986. *The Genius of China.* Simon and Schuster, New York.

Testart, A. 1982. "The Significance of Food Storage Among Hunter-Gatherers: Residence Patterns, Population Densities, and Social Inequalities." *Current Anthropology* 23:523–537.

Thomas, B. 1986. "Was There an Energy Crisis in Great Britain in the 17th Century?" *Explorations in Economic History* 23:124–152.

Thompson, T. M., J. K. Elkins, J. H. Butler, and B. D. Hall. 1990. "Nitrous Oxide and Halocarbon Group." In W. Komhyr, ed., *Climate Monitoring and Diagnostics Laboratory Summary Report 1989,* U.S. Department of Commerce, Boulder, Colo., pp. 64–72.

Thomsen, C. J. 1836, *Ledetraad til nordisk oldkyndighed.* L. Mellers, Copenhagen.

Thomson, K. S. 1987. "How to Sit on a Horse." *American Scientist* 75:69–71.

Tillman, D. A. 1978. *Wood as an Energy Resource.* Academic Press, New York.

Tilton, J. E., and B. J. Skinner. 1987. "The Meaning of Resources." In D. J. McLaren and B. J. Skinner, eds., *Resources and World Development,* John Wiley & Sons, Chichester, pp. 13–27.

Titow, J. Z. 1969. *English Rural Society, 1200–1350.* George Allen and Unwin, London.

Tompkins, P. 1976. *Mysteries of Mexican Pyramids.* Harper & Row, New York.

Torr, G. 1964. *Ancient Ships.* Argonaut Publishers, Chicago.

Torrey, V. 1976. *Wind-Catchers. American Windmills of Yesterday and Tomorrow.* Stephen Greene Press, Brattleboro, Vt.

Toth, N. 1987. "The First Technology." *Scientific American* 256(4):112–121.

Turner, B. L., II. 1990. "The Rise and Fall of Population and Agriculture in the Central Maya Lowlands: 300 B.C. to Present." In L. F. Newman, ed., *Hunger in History,* Blackwell, Oxford, pp. 78–211.

Turner, B. L., II, W. C. Clark, R. W. Kates, J. F. Richards, J. T. Matthews, and W. B. Meyer, eds. 1990. *The Earth as Transformed by Human Action.* Cambridge University Press, Cambridge.

UNESCO (United Nations Educational, Scientific, and Cultural Organization). 1960–. *Statistical Yearbook*. UNESCO, Paris.

Unger, R. 1984. "Energy Sources for the Dutch Golden Age." *Research in Economic History* 9:221–253.

UNO (United Nations Organization). 1956. "World Energy Requirements in 1975 and 2000." In *Proceedings of the International Conference on the Peaceful Uses of Atomic Energy*, Vol. 1, UNO, New York, pp. 3–33.

———. 1976. *World Energy Supplies 1950–1974*. UNO, New York.

———. 1982–. *Energy Statistics Yearbook*. UNO, New York.

———. 1992. *Statistical Yearbook*. UNO, New York.

Urbanski, T. 1967. *Chemistry and Technology of Explosives*. Pergamon Press, Oxford.

U.S. Bureau of the Census. 1950–. *Statistical Abstract of the United States*. U.S. Department of Commerce, Washington, D.C.

———. 1975. *Historical Statistics of the United States: Colonial Times to 1970*. U.S. Government Printing Office, Washington, D.C.

USDA (U.S. Department of Agriculture). 1955. *Wheat: Average Yields and Production by States, 1866–1943*. USDA, Washington, D.C.

———. 1959. *Changes in Farm Production and Efficiency*. USDA, Washington, D.C.

Usher, A. P. 1954. *A History of Mechanical Inventions*. Harvard University Press, Cambridge, Mass.

Van Beek, G. W. 1987. "Arches and Vaults in the Ancient Near East." *Scientific American* 257:(2):96–103.

van Duijn, J. J. 1983. *The Long Wave in Economic Life*. Allen & Unwin, Boston.

Vasko, T., R. Ayres, and L. Fontvieille, eds. 1990. *Life Cycles and Long Waves*. Springer-Verlag, Berlin.

Vavilov, N. I. 1951. *Origin, Variation, Immunity and Breeding of Cultivated Plants*. Chronica Botanica, Waltham, Mass.

Ville, S. P. 1990. *Transport and the Development of European Economy, 1750–1918*. Macmillan, London.

Villiers, G. 1976. *The British Heavy Horse*. Barrie & Jenkins, London.

Vogel, H. U. 1993. "The Great Well of China." *Scientific American* 268(6):116–121.

von Bertalanffy, L. 1968. *General System Theory*. George Braziller, New York.

von Braun, W., and F. I. Ordway. 1975. *History of Rocketry and Space Travel*. Thomas Y. Crowell, New York.

von Hippel, F. N., B. G. Levi, T. A. Postol, and W. H. Daugherty. 1988. "Civilian Casualties from Counterforce Attacks." *Scientific American* 259(3):36–42.

von Liebig, J. 1843. *Die Chemie in ihrer Anwendung auf Agricultur und Physiologie*. F. Vieweg, Braunschweig.

von Tunzelmann, G. N. 1978. *Steam Power and British Industrialization to 1860*. Clarendon Press, Oxford.

Waldron, A. 1990. *The Great Wall of China*. Cambridge University Press, Cambridge.

Waterbury, J. 1979. *Hydropolitics of the Nile Valley*. Syracuse University Press, Syracuse.

Watkins, G. 1967. "Steam Power—An Illustrated Guide." *Industrial Archaeology* 4(2):81–110.

Watt, B. K., and A. L. Merrill. 1975. *Handbook of the Nutritional Contents of Foods*. Dover Publications, New York.

Weast, R. C., ed. 1992. *RCRC Handbook of Chemistry and Physics*. CRC Press, Boca Raton, Fla.

Welsch, R. L. 1980. "No Fuel Like an Old Fuel." *Natural History* 89(11):76–81.

Wenke, R. J. 1989. "Egypt: Origins of Complex Societies." *Annual Review of Anthropology* 18:129–155.

Wesley, J. P. 1974. *Ecophysics*. Charles C. Thomas, Springfield, Ill.

White, D. A. 1987. "Conventional Oil and Gas Resources." In D. J. McLaren and B. J. Skinner, eds., *Resources and World Development*, John Wiley, Chichester, pp. 113–128.

White, K. D. 1970. *Roman Farming*. Thames & Hudson, London.

———. 1984. *Greek and Roman Technology*. Cornell University Press, Ithaca, N.Y.

White, L. 1978. *Medieval Religion and Technology*. University of California Press, Berkeley.

White, L. A. 1943. "Energy and the Evolution of Culture." *American Anthropologist* 45:335–356.

White, R. 1986. *Dark Caves, Bright Visions: Life in Ice Age Europe*. W. W. Norton, New York.

Whiten, A., and E. M. Widdowson, eds. 1992. *Foraging Strategies and Natural Diet of Monkeys, Apes and Humans*. Clarendon Press, Oxford.

Whitmore, T. M., B. L. Turner II, D. L. Johnson, R. W. Kates, and T. R. Gottschang. 1990. "Long-term Population Change." In B. L. Turner II, W. C. Clark, R. W. Kates, J. F. Richards, J. T. Matthews, and W. B. Meyer, eds., *The Earth as Transformed by Human Action*, Cambridge University Press, New York, pp. 25–39.

Whitt, F. R., and D. G. Wilson. 1982. *Bicycling Science*. MIT Press, Cambridge, Mass.

Williams, D.S.D., ed. 1972. *The Modern Diesel*. Newnes-Butterworths, London.

Williams, T. I. 1987. *The History of Invention: From Stone Axes to Silicon Chips*. Facts on File, New York.

Williamson, J. G. 1982. "Was the Industrial Revolution Worth It? Disamenities and Death in 19th Century British Towns." *Explorations in Economic History* 19:221–245.

Wilson, C. 1990. *The Gothic Cathedral*. Thames and Hudson, London.

Wilson, P. N. 1956. *Watermills: An Introduction*. Times Printing, Mexborough.

Winterhalder, B., R. Larsen, and R. B. Thomas. 1974. "Dung as an Essential Resource in a Highland Peruvian Community." *Human Ecology* 2:89–104.

Wolff, A. R. 1900. *The Windmill as Prime Mover*. John Wiley, New York.

Womack, J. P., D. T. Jones, and D. Roos. 1991. *The Machine that Changed the World*. Harper Perennial, New York.

Wood, W. 1922. *All Afloat*. Glasgow, Brook & Co., Toronto.

World Resources Institute. 1990. *World Resources 1990–91*. Oxford University Press, New York.

———. 1992. *World Resources 1992–93*. Oxford University Press, New York.

Wright, G. A. 1971. "Origins of Food Production in Southwestern Asia: A Survey of Ideas." *Current Anthropology* 12:447–477.

Wright, O. 1953. *How We Invented the Airplane*. David McKay, New York.

Wu, K. C. 1982. *The Chinese Heritage*. Crown Publishers, New York.

Yates, R. S. 1990. "War, Food Shortages, and Relief Measures in Early China." In L. F. Newman, ed., *Hunger in History*, Basil Blackwell, Oxford, pp. 147–177.

Yesner, D. R. 1980. "Maritime Hunter-Gatherers: Ecology and Prehistory." *Current Anthropology* 21:727–750.

Yoffee, N., and G. L. Cowgill, eds. 1988. *The Collapse of Ancient States and Civilizations,* University of Arizona Press, Tucson, Ariz.

Ziemke, E. F. 1968. *The Battle for Berlin: End of the Third Reich.* Ballantine, New York.

Zvelebil, M. 1986. "Postglacial Foraging in the Forests of Europe." *Scientific American* 254(5):104–115.

About the Book and Author

Every human activity entails the conversion of energy. Changes in the fundamental sources of energy, and in the use of energy sources, are a basic dimension of the evolution of society. Our appreciation of the significance of these processes is essential to a fuller understanding of world history.

Vaclav Smil offers a comprehensive look at the role of energy in world history, ranging from human muscle-power in foraging societies and animal-power in traditional farming to preindustrial hydraulic techniques and modern fossil-fueled civilization. The book combines a vast historical sweep with cross-cultural comparisons and is enhanced by illustrations and accessible quantitative material. Students and general readers alike will gain an understanding of energy's fundamental role in human progress.

Smil illuminates the role played by various means of harnessing energy in different societies and provides new insights by explaining the impact and limitations of these fundamental physical inputs—whether it is in the cultivation of crops, smelting of metals, waging of war, or the mass production of goods. While examining the energetic foundations of historical changes, *Energy in World History* avoids simplistic, deterministic views of energy needs and recognizes the complex interplay of physical and social realities.

Vaclav Smil teaches at the University of Manitoba. He has written thirteen books, including *Energy, Food, Environment: Realities, Myths, Options; General Energetics;* and *Global Ecology, Environmental Change, and Social Flexibility.*

Index